Mathematics of Life

By the same author

Concepts of Modern Mathematics
Game, Set and Math
The Problems of Mathematics
Does God Play Dice?
Another Fine Math You've Got Me into
Fearful Symmetry (with Martin Golubitsky)
Nature's Numbers
From Here to Infinity
The Magical Maze
Life's Other Secret
Flatterland
What Shape Is a Snowflake?
The Annotated Flatland
Math Hysteria
The Mayor of Uglyville's Dilemma
Letters to a Young Mathematician
Why Beauty Is Truth
How to Cut a Cake
Taming the Infinite: The Story of Mathematics
Professor Stewart's Cabinet of Mathematical Curiosities
Professor Stewart's Hoard of Mathematical Treasures
Cows in the Maze

with Terry Pratchett and Jack Cohen
The Science of Discworld
The Science of Discworld II: the Globe
The Science of Discworld III: Darwin's Watch

with Jack Cohen
The Collapse of Chaos
Figments of Reality
Evolving the Alien/What Does a Martian Look Like?
Wheelers (science fiction)
Heaven (science fiction)

Mathematics of Life

Unlocking the Secrets of Existence

Ian Stewart

PROFILE BOOKS

First published in Great Britain in 2011 by
PROFILE BOOKS LTD
3A Exmouth House
Pine Street
London EC1R 0JH
www.profilebooks.com

A CIP catalogue record for this book is available from the British Library.

ISBN 978 184668 198 1
eISBN 978 184765 350 5
Export ISBN 978 184668 550 7
Text design by sue@lambledesign.demon.co.uk
Typeset in Stone Serif by Data Standards Ltd, Frome, Somerset
Printed and bound in Britain by Clays Ltd, Bungay, Suffolk

The paper this book is printed on is certified by the © 1996 Forest Stewardship
Council A.C. (FSC). It is ancient-forest friendly. The printer holds FSC chain of
custody SGS-COC-2061.

FSC
Mixed Sources
Product group from well-managed
forests and other controlled sources
Cert no. SGS-COC-2061
www.fsc.org
© 1996 Forest Stewardship Council

Contents

Preface // vii

1   Mathematics and Biology // 1
2   Creatures Small and Smaller // 13
3   Long List of Life // 30
4   Florally Finding Fibonacci // 38
5   The Origin of Species // 56
6   In a Monastery Garden // 77
7   The Molecule of Life // 91
8   The Book of Life // 111
9   Taxonomist, Taxonomist, Spare that Tree // 124
10 Virus from the Fourth Dimension // 138
11 Hidden Wiring // 158
12 Knots and Folds // 181
13 Spots and Stripes // 198
14 Lizard Games // 213
15 Networking Opportunities // 246
16 The Paradox of the Plankton // 258
17 What is Life? // 275
18 Is Anybody Out There? // 289
19 The Sixth Revolution // 317

Notes // 320
Acknowledgements // 335
Index // 337

Preface

Mathematical theory and practice have always gone hand in hand, from the time primitive humans scratched marks on bones to record the phases of the Moon to the current search for the Higgs boson using the Large Hadron Collider. Isaac Newton's calculus informed us about the heavens, and over the past three centuries its successors have opened up the whole of mathematical physics: heat, light, sound, fluid mechanics, and later relativity and quantum theory. Mathematical thinking has become the central paradigm of the physical sciences.

Until very recently, the life sciences were different. There, mathematics was at best a servant. It was used to perform routine calculations and to test the significance of statistical patterns in data. It didn't contribute much conceptual insight or understanding. It didn't inspire great theories or great experiments. Most of the time, it might as well not have existed.

Today, this picture is changing. Modern discoveries in biology have opened up a host of important questions, and many of them are unlikely to be answered without significant mathematical input. The variety of mathematical ideas now being used in the life sciences is enormous, and the demands of biology are stimulating the creation of entirely new mathematics, specifically aimed at living processes. Today's mathematicians and biologists are working together on some of the most difficult scientific problems that the human race has ever tackled – including the nature and origin of life itself.

Biology will be the great mathematical frontier of the twenty-first century.

Mathematics of Life celebrates the rich variety of connections between mathematics and biology that already exist, from the Human Genome Project, through the structure of viruses and the organisation of the cell, to the form and behaviour of entire organisms and their interaction in the global ecosystem. It will also show how mathematics can shed new light on difficult issues concerning evolution, where many important processes take too long to observe, or happened hundreds of millions of years ago and have left only cryptic traces.

Initially, biology was about plants and animals. Then it was about cells. Now it is mostly about complex molecules. To reflect these changes in scientific thinking about the enigma of life, the book starts from the everyday human level, and follows the historical path that led biologists to·focus ever more sharply on the microscopic structure of living creatures, culminating in DNA, the 'molecule of life'.

Most of the material discussed in the first third of the book is therefore about biology. However, mathematics makes an early appearance, tracing questions about the geometry of plants from Victorian times to the present day, to illustrate how new mathematical ideas have been motivated by biology. Once the biological background has been established, mathematics comes to centre stage as we build up from the atomic scale, back to the level with which we feel most comfortable, the one on which we all live. The world of grass, trees, sheep, cows, cats, dogs ... and people.

The mathematics involved is far-ranging: probability, dynamics, chaos theory, symmetry, networks, mechanics, elasticity – even knot theory. Most of the applications discussed here are to mainstream mathematical biology: the structure and function of the complex molecules that co-ordinate the complex processes of life, the shapes of viruses, the evolutionary games that led to the huge diversity of life on this planet and are still happening today, the workings of the nervous system and the brain, the dynamics of ecosystems. I've also included chapters on the nature of life and the possible existence of alien life forms.

The interaction between mathematics and biology is one of the hottest areas of science. It has already come a long way in a very short time. Only the future will show just how far it can go. But one thing I guarantee: it's going to be an exciting ride.

Ian Stewart
Coventry, September 2010

1   Mathematics and Biology

· ·

Biology used to be about plants, animals and insects, but five great revolutions have changed the way scientists think about life.

A sixth is on its way.

The first five revolutions were the invention of the microscope, the systematic classification of the planet's living creatures, the theory of evolution, the discovery of the gene, and the discovery of the structure of DNA. Let's look at them in turn, before moving on to my sixth, more contentious, revolution.

· · · · · · · · · · · · · · · ·

The Microscope

The first biological revolution happened 300 years ago, when the invention of the microscope opened our eyes to the astonishing complexity of life on the smallest scales. More precisely, it opened up the complexity of life to observation by our eyes, by providing a new instrument to augment our unaided senses.

The invention of the microscope led to the discovery that individual organisms have an amazing internal complexity. One of the first big surprises was that living creatures are made from cells – tiny bags of chemicals enclosed in a membrane that lets some of the chemicals pass in or out. Some organisms consist of a single cell, but even those are surprisingly complicated, because a cell is an entire chemical system, not something simple and straightforward. Many organisms are made from a gigantic number of cells: your body contains roughly 75 trillion of them. Each cell is a tiny biological machine with its own genetic machinery which can cause it to reproduce, or die. Cells come in more than 200 types – muscle cells, nerve cells, blood cells, and so on.

Cells were discovered very soon after microscopes were invented: once you can look at an organism under high magnification, you can't miss them.

. . . .

Classification

The second revlution was started by Carl Linnaeus, a Swedish botanist, doctor and zoologist. In 1735, his epic work *Systema Naturae* appeared. Its full title in English is 'The system of nature through the three kingdoms of nature, according to classes, orders, genera and species, with characters, differences, synonyms, places'. Linnaeus was so interested in the natural world that he decided it needed to be catalogued. All of it. The first edition of his catalogue was just 11 pages long; the 13th and last ran to 3,000 pages. Linnaeus made it clear that he was not trying to uncover some kind of hidden natural order; he was just trying to organise what was there, in a systematic and structured manner. His chosen structure was to classify natural objects in a five-stage subdivision: kingdom, class, order, genus, species. His three kingdoms were animals, plants and minerals. He founded the science of taxonomy: the classification of living creatures into related groups.

Minerals are no longer classified along Linnaean lines, and the details of his system have been modified for plants and animals. Recently several alternative systems have been advocated, but none has yet been widely adopted. Linnaeus appreciated that a systematic classification of living things is vital to science, and he put that idea into practice. He made the occasional mistake: initially he classified whales as fish. But by the 10th edition of *Systema Naturae*, published in 1758, an ichthyologist friend had put him right, and whales were mammals.

The best-known and most useful feature of the Linnaean system is the use of double-barrelled names such as *Homo sapiens*, *Felis catus*, *Turdus merula* and *Quercus robur* – species of human, cat, blackbird and an oak tree.[1] The importance of classification is not just to make a list, or to introduce fancy Latinised names to show how clever you are, but to make logical, clear-cut distinctions among the many creatures that exist. Common names, such as 'blackbird', don't do the trick: do you mean the common blackbird, the grey-winged blackbird, the Indian blackbird, the Tibetan blackbird, the white-collared blackbird, or one of the 26 species of New World blackbird? But the Linnaean *Turdus merula* refers uniquely to the common blackbird, and there's no chance of confusion.

· · · · · · · · · · · · · · · ·

Evolution

The third revolution had been simmering for some time, but it boiled over in 1859 when Darwin published *The Origin of Species*. The book eventually ran to six editions, and it ranks as one of the truly great scientific works of all time, bearing comparison with the works of Galileo, Copernicus, Newton and Einstein in the physical sciences. In the *Origin*, Darwin proposed a new vision of the source of life's diversity.

The prevailing belief in his day, among scientists as much as lay folk, was that each separate species had been created individually by God as part of the overall act of creating the universe. In this view, species could not change over time: a sheep was, is and always will be a sheep; a dog was, is and always will be a dog. But as Darwin contemplated the scientific evidence, much of which he had amassed on his own travels, he found this comfortable picture becoming less and less tenable.

Pigeon fanciers knew that deliberate breeding could produce wildly different types of pigeon. The same went for cows, dogs and indeed all domesticated animals. Now, that mechanism for change required human intervention. The animals didn't change 'of their own accord': they had to be chosen – *selected* – with great care, by someone following a plan. But Darwin realised that unaided nature could, in principle, produce similar changes through competition for resources. When times were hard, those animals that were better able to survive would be the ones that lived long enough to produce the next generation, and this new generation would be slightly better adapted to the environment.

Such changes, Darwin felt, would be much more gradual than those imposed by human breeders, but a changing environment could, over a long period of time, cause some of the organisms in a species to develop markedly different forms and habits. He saw this process as the slow accumulation of myriad tiny changes. His background in geology made him acutely aware that the planet had been around for vast aeons of time, so lack of time was not a problem. Even extraordinarily slow changes could eventually become very significant.

He called this process 'natural selection'. Today we call it 'evolution', a word that Darwin didn't use – although the final word in *The Origin of Species* is 'evolved'. The evidence in favour of evolution is so extensive, and comes from so many independent sources, that biology now makes no sense without it. Today, almost all biologists (and most scientists, whatever their field of research) find the evidence that evolution has been the dominant mechanism behind the diversity of today's species to be overwhelming. But how evolution works is another matter entirely, and much remains to be understood.

· · · · · · · · · · · · · · · ·

Genetics

The fourth revolution was Gregor Mendel's discovery of genes, which was published in 1865 but not appreciated for another fifty years.

Observable features of organisms, such as colour, size, texture and shape, are known as characters (or characteristics or traits). Darwin had no idea how characters were transmitted from parent to offspring, though several distinct lines of reasoning led him to infer that this must happen. In fact, the transmission mechanism was already under investigation when he wrote the *Origin*, but he didn't know that. It would have had a major impact on his thinking.

For seven years around 1860, the Austrian priest Gregor Mendel bred pea plants – 29,000 of them – and counted how many displayed particular characters in each generation. Did they produce yellow or green peas? Were the peas smooth or wrinkly? Mendel's observations turned up some curious mathematical patterns, and he became convinced that inside every living organism there are 'factors', now called genes, that somehow determine many features of the organism itself. These factors are inherited from previous generations, and in sexual species they arise in pairs: one from the 'father' (the male organ of the plant) and one from the 'mother' (the female organ). Each factor can occur in several distinct forms.

The random mixing of these 'alleles' – genetic alternatives – creates the patterns in the numbers.

Initially, the physical form of Mendel's factors was a complete mystery; their existence was inferred indirectly from the mathematical patterns – the proportions of plants in successive generations that possessed particular combinations of features.

• • • • • • • • • • • • • • • •

The structure of DNA

Revolution number five was more straightforward, and like the first, it was triggered by the invention of a new experimental technique. This time the technique was X-ray diffraction, which allows biochemists to work out the structure of complex, biologically important molecules. In effect, it provides a 'microscope' that can reveal the positions of individual atoms in a molecule.

In the 1950s Francis Crick and James Watson began to think about the structure of a complex molecule found almost universally in living creatures: deoxyribose nucleic acid, known universally by its initials, DNA. Crick, who was British, had trained as a physicist, but became terminally bored while writing a PhD on how to measure the viscosity of water at high temperatures, and in 1947 he moved into biochemistry. Watson was an American whose first degree was in zoology; he became interested in a type of virus that infects bacteria, known as a bacteriophage ('bacterium-eater'). His big project was to understand the physical nature of the gene – its molecular structure.

At that time, it was known that genes resided in regions of the cell called chromosomes, and that the main constituents of genes were proteins and DNA. The conventional wisdom among biologists was that organisms could reproduce because the genes were proteins, capable of copying themselves. DNA, in contrast, was widely considered to be a 'stupid tetranucleotide' whose sole function was to act as scaffolding, so that the proteins could be held together.

However, there was already some evidence that DNA is the molecule from which genes are formed, which immediately raised a crucial question: what does the DNA molecule look like? How are its component atoms arranged?

Watson ended up working with Crick. They based their analysis of DNA on some crucial X-ray diffraction experiments carried out by others (notably Maurice Wilkins and Rosalind Franklin), homed in on a few key facts, and started building models in the literal sense, by fitting together pieces of card or metal shaped like simple molecules that were known to be part of DNA. This exercise led them to propose the now-famous double helix structure: DNA is two-stranded, like two intertwined spiral staircases. Each strand (staircase) carries a series of bases, which are four different molecules: adenine (A), cytosine (C), guanine (G) and thymine (T). These come in linked pairs: an A on one strand is always joined to a T on the other; a C on one strand is always joined to a G on the other.

Crick and Watson published their proposal in the scientific journal *Nature* in 1953. It begins: 'We wish to suggest a structure for the salt of deoxyribose nucleic acid (D.N.A.). This structure has novel features which are of considerable biological interest.' Near the end, they write: 'It has not escaped our notice that the specific pairing we have postulated [A with T, C with G] immediately suggests a possible copying mechanism for the genetic material.'[2]

The basic idea here is simple: the sequence of bases on just one of the two strands determines the entire structure. On the other strand, the sequence is given by the complementary bases to those on the first strand – swap A and T, and swap C and G. If you could pull DNA apart into its two strands, each of them would contain the necessary 'information' to reconstruct the other. So all you have to do is make two complementary strands, and fit the pairs back together to get two perfect copies of the original.

Crick and Watson's suggestion for the structure of DNA, based on little more than some crucial hints from experiment and a lot of fiddling with models, turned out to be correct. So did the copying mechanism, which was so speculative that they did not spell it out in the *Nature* paper in case it turned out to be wrong. However, you can't just pull two intertwined helices apart, so some quite complicated mechanisms are needed to achieve this duplication. What they were lay far in the future.

At a stroke, attention in biology turned to the molecular structure of key substances: DNA, proteins and associated molecules. University biology departments fired or retired botanists, zoologists and taxonomists – anyone who actually worked with *entire animals* was completely out of date. Molecules were the coming thing. And they were, and they did. And biology has never been the same since. Crick and Watson had found 'the secret of life', as Crick bragged in the *Eagle* (a pub in Benet Street, Cambridge) a few days before they found the correct structure.

Many major new developments have followed from Crick and Watson's breakthrough. The science behind them is often highly innovative, but the point of view has changed only incrementally from what it was in Crick and Watson's day, so these more recent scientific advances, dramatic though they may be, do not constitute genuine revolutions. For example, in 2006 the Human Genome Project succeeded in listing the entire genetic sequence of a human being – three billion units of genetic information.[3] This has revolutionary implications: for one thing, it opens up entirely new advances in medicine. Biology has become the most exciting scientific frontier of the twenty-first century, promising huge advances in medicine and agriculture, as well as a deep understanding of the nature of life itself. But there is a clear path linking all of this to the original discovery of the structure of DNA.

. . . .

These, then, are my five revolutions.

The gaps between them, allowing (in Mendel's case) for the time it took before anyone noticed, are roughly 50, 100, 50 and 50 years. The fifth happened just over 50 years ago. The pace of change in the world is accelerating, so a sixth revolution in biology seems overdue. I believe that it has already arrived. The nature of life is not just a question for biochemistry – many other areas of science have major roles in explaining what makes living creatures live. What unites them all, opening up entirely new vistas, is my sixth biological revolution: mathematics.

. . . .

Mathematics has been with us for thousands of years; the ancient Babylonians could solve quadratic equations 4,000 years ago. Biologists have been using mathematical techniques, especially statistics, for more than a century. So it might seem unreasonable to refer to a 'revolution'. But what I have in mind – what is happening as I write – goes much further. The mathematical way of thinking is becoming a standard piece of kit in the biological toolbox: not just a way to analyse data about living creatures, but a method for understanding them.

What mathematics is, and how useful it is, are widely misunderstood. It is not solely about numbers, 'doing sums' as we were taught in school – that's arithmetic. Even when you add in algebra, trigonometry, geometry and various more modern topics such as matrices, what we learn at school is a tiny, limited part of a vast enterprise. To call it one-tenth of one per cent would be generous. And the mathematics we learn at school is in many ways unrepresentative of the whole, just as playing scales on a piano falls short of being real music, and woefully short of *composing* music. People often think that mathematics was all invented (or discovered) long ago, but new mathematics is coming into being at an impressive rate. A million pages a year is a conservative estimate, and that's a million pages of new ideas, not just variations on routine calculations.

Numbers are basic to mathematics, just as scales are basic to music, but the subject matter of mathematics is much broader: shapes, logic, processes ... anything that has structure or pattern. We can also include uncertainty, which might seem to be the absence of pattern, but the early statisticians discovered that even random events have their own patterns, on average and in the long run. One of the remarkable features of the mathematics now being used in biology is its variety; another is its novelty. Much of it is less than 50 years old and some of it was invented last week. It ranges from knot theory to game theory, from differential equations to symmetry groups. A lot of it uses ideas that most of us have never encountered, and probably wouldn't recognise as mathematics if we did. It is changing how we *think* about biology, not just the results we obtain.

This approach is old hat in the physical sciences, which rely heavily on mathematics; in fact, the development of those two areas has gone hand in hand for thousands of years. Until recently, biology was – or seemed – different. Traditionally, biology was the branch of science recommended to students who preferred to avoid mathematics if at all possible. You can study the life cycle of a butterfly without doing any sums. There are still no fundamental mathematical equations for biology, equivalents of Newton's law of gravitation. We don't calculate the evolutionary trajectory of a fish by applying Darwin's equations. But there is mathematics aplenty in today's biology, and it is becoming ever harder to avoid it. It just doesn't mimic the way mathematics is used in physics. It's different, it has its own special quality. And increasingly, much of it is motivated by the needs of biologists, which are no longer as cosy as watching butterflies.

The application of mathematics to biology depends on new apparatus, most obviously the computer. It also depends on new mental apparatus: mathematical techniques, some specially tailored to the needs of biology, others that arose for different reasons but turn out to have important biological implications. Mathematics provides a new point of view, addressing not just the ingredients for life, but the processes that use those ingredients.

I believe that the sixth revolution in biology is already under way, and it is to apply mathematical insight to biological processes. My aim here is to show how the techniques and viewpoints of mathematics are helping us to understand not just what life is made from, but how it works, on every scale from molecules to the entire planet – and possibly beyond.

• • • • • • • • • • • •

Until recently, most biologists doubted that mathematics would ever have much to tell us about life. Living creatures seemed too versatile, too flexible, to conform to any rigid mathematical formalism (hence the Harvard law of animal behaviour: 'experimental animals, under carefully controlled laboratory conditions, do what they damned well please'). Mathematical tools such as statistics had their place, of course, but mathematics was purely a servant, unlikely to have a significant effect on mainstream biological thinking. Mavericks such as D'Arcy Wentworth Thompson, whose book *On Growth and Form* catalogued numerous mathematical patterns – or alleged patterns – in living creatures, were ignored or dismissed. They were at best a sideshow, at worst, nonsense. After all, Thompson's book was first published in 1917, forty years before the structure of DNA became known, and he said very little about evolution, except to criticise what he saw as a tendency to fit the story to whatever facts happened to be available. More recent critics of a narrow molecular view of biology, such as the American evolutionary biologist Richard Lewontin, also got short shrift from mainstream biology. The genome was considered to be 'the information needed to specify an organism', and it was pretty obvious that once we knew that, then in principle we would know everything.

However, as biologists overcame the huge difficulties involved in deriving genetic sequences, and in working out the functions of genes and proteins – what they actually *did* in the organism – the true depths of the problem of life became ever more apparent. Listing the proteins that make up a cat does not tell us everything we want to know about cats. It doesn't tell us everything even for more lowly creatures such as bacteria.

There is no question that a creature's genome is fundamental to its form and behaviour, but the 'information' in the genome no more tells us everything about the creature than a list of components tells us how to build furniture from a flat-pack. In fact, the gulf between a living creature and its genome is far wider than that between furniture and a list of boards, screws and washers. For example, over the past few years it has also become clear that 'epigenetic' information, not written in DNA, and possibly not 'coded' in any obvious symbolic fashion, is also vital to life on Earth. Most of us who have assembled flat-packs have also required knowledge that is not included in the instructions.

Lists of ingredients are not enough to understand biology, because what really matters is how those ingredients are used – the processes that they undergo in a living creature. And the best tool we possess for finding out what processes do is mathematics. Over the past half-century or so, new mathematical discoveries have opened up a realm of rich and surprising behaviour, revealing that apparently simple processes can do astonishingly complex things. As a result, the belief that mathematics is too simple and too well behaved to provide insights into the complexity of living creatures has become very difficult to defend. Instead, attention has been focused on finding ways to exploit the power of mathematics to provide genuine insights into biology.

Some of these developments use mathematics as a tool to help with the scientific techniques that biologists use. Such applications have been around ever since physicists developed the science of optics and manufacturers used it to improve the design of microscopes. An example today is 'bioinformatics', the methods involved in the storage and manipulation of gigantic data sets in computers. Listing a genome is not enough: you have to be able to find what you're looking for in the list, compare it with other items of information on other lists, and so on. When the list contains three billion items of information (and that's just the code, let alone everything we know about what it does), that's a non-trivial issue. Most computer technology relies on a heavy dose of hidden mathematics, and bioinformatics is no exception.

That's worthy, useful, necessary ... but not, in the present context, inspiring. The role of mathematics ought to be more creative. And so it is. Mathematics is being used not just to help biologists manage their data, or improve their instruments, but on a deeper level: to provide significant insights into the science itself, to help explain how life works. Over the past ten years there has been a massive growth in 'biomathematics' – mathematical biology. All around the globe new research institutes and centres devoted to this subject have sprung into existence, to such an extent that the people setting them up are having difficulty in finding enough qualified staff. Though still not a part of the biological mainstream, biomathematics is claiming its rightful place among the host of techniques and points of view that are necessary if we are to understand how life evolved, how it works and how organisms relate to their environment.

Ten or twenty years ago, the claim that mathematics could play a significant role in biology largely fell on deaf ears. Today, that particular battle is mostly won – as the rapid growth of specialist research centres demonstrates. It is no longer necessary to try to persuade biologists that mathematics might be useful to them. Many of them still have no wish to use it themselves, except when it has been neatly packaged into computer software, but they do not object if others do. A mathematician can be a useful addition to the research team. A few biologists still resist the importation of mathematics into their subject and would robustly deny most of what I've just written, but that's fast becoming an outmoded reflex, and their influence is dwindling.

By the same token, mathematicians have learned that the only effective way to apply their subject to biology is to find out what biologists want to know, and to adapt their techniques accordingly. Biomathematics is not merely a new application for existing mathematical methods. You can't just pull an established mathematical technique off the shelf and put it to use: it has to be tailored to fit the question. Biology requires – indeed demands – entirely new mathematical concepts and techniques, and it raises new and fascinating problems for mathematical research.

If the main driving force behind new mathematics in the twentieth century was the physical sciences, in the twenty-first century it will be the life sciences. As a mathematician, I find this prospect exciting and enticing. Mathematicians like nothing better than a rich source of new questions. Biologists, rightly, will be impressed only by the answers.

2   Creatures Small and Smaller

. . . .

If human eyesight had been better, we might never have
experienced the first revolution, when we noticed the hidden
wonders of life. Our poor eyesight inspired a simple piece of
technology – the lens. Unexpectedly, this practical aid to our
everyday activities spun off two types of scientific instrument: the
telescope and the microscope. These opened up the vast reaches of
the cosmos and the intricate small-scale world of living creatures.

Unaided, the human eye sees the world on a human scale:
people, houses, animals, plants, rocks, rivers, cups, knives ... Even
the larger features of our environment – mountains, lakes – we
perceive as monolithic objects. From a distance, a mountain looks
much like a rock, one that comes to a point at the top. By the time
we are close enough to see how much more there is to a mountain,
we have lost sight of the mountain. Instead, we see a complex
arrangement of streams, scattered rocks, moss, precipices, ravines,
snow and ice.

The word 'grasp' gives away the whole game. On a human scale,
the world consists of what we can pick up with our hands. On this
level, the Moon, a cow and a flea seem to be on a par with one
another. Agreed, we can't grasp the Moon, but we can cover it with
a thumb held at arm's length. We can't pick up a cow, but we can
put a ring through its nose and lead it on a rope. (I use 'we' in the
time-honoured sense of 'some of us can'.) The main problem in
grasping a flea, ironically, is that it's too small to offer a good grip –
and it jumps. But broadly speaking, on a human scale all objects are
on much the same footing. We give them a name, and we imagine
that by naming them we have captured their essence. The Moon is

a shining, mottled disc. A cow walks on four legs and gives milk. A flea bites, jumps and is a nuisance.

As soon as we progress beyond the unaided human eye, with little more than a polished lump of glass to assist us, our simple, comfortable picture of the world changes. Through his telescope Galileo saw spots on the Sun, mountains on the Moon, phases of Venus, and four tiny specks of light passing to and fro across the orange disc of the planet Jupiter. He deduced – could scarcely fail to deduce, as soon as he put his mind to it – that the Sun and Moon are not unblemished spheres, Venus revolves around the Sun, and the Earth is not a fixed centre around which the rest of the universe revolves.

The religious authorities of the time, who considered themselves to be custodians of the truth, were aghast. Galileo managed to escape the horrific penalties that were often employed to enforce the official view of truth, but at his trial for heresy in 1633 he was forced to deny his own deductions from the evidence that he had seen through his telescope. The authorities of the day did not dispute the evidence. They simply told Galileo to ignore it, and stop writing about it. I'm inclined to believe that they acted like this not because they were religious, but because they were authorities.

So Galileo recanted, though allegedly muttering under his breath 'even so, it moves'. And the Earth continued to move round the Sun, whatever the Church believed and whatever Galileo was publicly obliged to assert. The scientific evidence eventually prevailed, but by the time Pope John Paul II apologised for how Galileo had been treated, science had put men on the Moon.

If a humble telescope could cause such ructions, merely by revealing things that were *there*, what about the microscope? That opened up the internal world of very small things – in particular, living creatures. The potential for heretical ideas was far greater than anything that astronomy could inspire. Yet curiously, the religious authorities viewed this even more revolutionary development with apparent equanimity, even though the new evidence now made available to the human eye would totally change our ideas about the world and our place in it. I suspect that the authorities simply didn't grasp the microscope's potential. The wonders it revealed did not, initially, appear to conflict with scripture. The Church, taking a positive religious message, believed the microscope was merely showing us the hidden marvels of God's creation. A pity they didn't think the same about Galileo.

In fact, the microscope was far from innocuous. It quickly revealed that our world is not what it seems. It does not function solely on the human level, it was not made *for* humans; everything that humans had been taking for granted about plants and animals was up for grabs, and most of it was wrong. Even those things, like cats and cows and trees, that *do* seem to function on a human level ... don't.

On the human level, a cow seems simple. You feed it grass, and it pays you back with milk. It's a trick whose secret is limited to cows and a few other mammals (most can't digest grass). You don't need to understand the details to exploit the process: it's a straightforward transformation from grass into milk, more like chemistry – or alchemy – than biology. It is, in its way, magic, but it's rational magic that works reliably. All you need is some grass, a cow and several generations of practical knowhow.

Seen through a microscope, though, it all gets more complicated. And the closer you look, the more complicated it gets. Milk is not a single substance, but a mixture of many. Grass is so complex that we still don't fully understand it. A cow's complexity is even greater. In particular, a cow (plus a bull) can make a new generation of baby cows. This is a simple thing on a human level, but inexpressibly complex on a microscopic level.

. . . .

Nearly three thousand years ago, the ancient Egyptians knew that a glass lens can make an object look bigger. Seneca, who tutored the Roman Emperor Nero, noticed that it is easier to read someone's writing if you look through a glass globe filled with water. Nero himself is said to have looked through an emerald to watch his gladiators fighting in the arena. By the ninth century, people were using 'reading stones' to assist their failing eyesight. These were polished lumps of clear glass, rounded on one side and flat on the other; you sat them on top of the document you were trying to read and looked through them. By the twelfth century, the Chinese had discovered that slices of smoky quartz can protect your eyes from the sun.

No one knows exactly when, where or by whom the first true

spectacles – a pair of lenses that you perch on your nose – were invented. One contender is Salvino D'Armati, who lived in Florence and may have invented spectacles around 1284. Another is a Dominican monk, Alessandro Spina, from Pisa. A third is Roger Bacon, whose 1235 (or earlier) book about the rainbow mentions using optical devices to read small letters from a distance. However, we have no idea what sort of device he had in mind. It may have been just a single crude lens.

Whoever should be given the credit, the first true spectacles were almost certainly invented in Italy between 1280 and 1300. They acted like a magnifying glass and corrected long-sightedness; it would be another 300 years before lenses able to correct short-sightedness were developed, in part because these are much harder to make. Johannes Kepler (astronomer, astrologer and mathematician) was the first to explain how convex and concave lenses corrected eyesight. Spectacles work better if the lenses are made from clear glass, without too many bubbles or impurities, and the precise shape of the lens is crucial. Lenses were (and still are) made by grinding glass using various types of abrasive material, which in Kepler's time were already being used by jewellers. So lens technology developed alongside other improvements.

In 1590 a Dutch spectacle manufacturer, Zaccharias Janssen, assisted by his son Hans, put several lenses inside a tube. When they looked through the tube, it made everything appear larger and nearer. This discovery led to two of the most important scientific instruments ever invented: the telescope and the microscope. The telescope brought the large, distant structures of the cosmos down to a human scale. The microscope did the exact opposite: it took the diminutive structures of Earthly objects, especially living creatures, and brought them up to the human scale.

By 1609 Galileo had improved these early telescopes, and through his still rather crude instruments he made discoveries that persuaded him that the Earth was not the centre of the universe. Within a century, astronomy had become a thriving area of science, and the secrets of the heavens, especially the laws of gravity, were there for the taking.

The telescope opened up astronomy because it made it possible for the human eye to see enormously distant, enormously large objects, such as planets. It took the exact opposite to open up biology: a device that allowed the human eye to see incredibly tiny objects that were right in front of people's noses. By a happy accident, the same basic technology – lenses – did this job too. The resulting device even has a similar name: the microscope.

The invention of the microscope had a very different effect from that of the telescope. It led to great strides in biology, but instead of clarifying the issues, many of those strides made things seem even more mysterious and miraculous. Instead of opening up the world of living creatures to human understanding, the microscope just made the puzzles seem even more difficult. Through even a low-powered microscope, little more than a single crude lens, living creatures took on new significance. And they were very, *very* complex.

So, while the telescope revealed deep simplicities in the cosmos, the microscope revealed previously unseen complexities in life. The same dichotomy between the simple and the complex has bedevilled the biological sciences ever since. Biologists, with some justification, argue that the life sciences are fundamentally harder than the physical sciences.

. . . .

A key figure in the development of the microscope was the Dutch tradesman and scientist Anton van Leeuwenhoek. He developed a way to make small, high-quality spheres of glass, and used them as lenses. Although a sphere is not the ideal shape for a lens, the quality of the glass compensated for the poor geometry, and van Leeuwenhoek's microscopes were surprisingly powerful. Using this new device, he became the first person to observe bacteria, yeast and microscopic creatures that dwelt in ponds. Under one of his microscopes, a drop of pondwater teemed with as much life as the Serengeti plain. He also discovered that blood was made from tiny disc-shaped objects, which flowed round the body in tiny tubes, capillaries.

Starting in 1673, van Leeuwenhoek published his discoveries in the *Philosophical Transactions*, a journal of the Royal Society in London. At first his work attracted favourable comment, but after three years he began to make claims that most scientists of the day found absurd: the discovery of 'animalcules'. These creatures, he said, flourished inside a single drop of water. The idea that there might exist living organisms so small that they were invisible to the naked eye seemed ludicrous, and at first van Leeuwenhoek's claim was met with derision.

The types of creature that van Leeuwenhoek discovered are nowadays known as protists. The best known protist must be 'the' amoeba, thanks to school biology; actually there are innumerable species of amoeba, some of which even have shells. So 'amoeba' has become a generic term for all such creatures (the technical term is 'amoeboid'). Amoebas were discovered in 1757 by August von Rosenhof, and initially they were referred to as 'Proteus animalcules', after the Greek god famed for his ability to change shape. The amoeba with the scientific name *Amoeba proteus* is the most familiar, mainly because it is also one of the largest, and so can easily be seen under a low-powered microscope (see Figure 1).

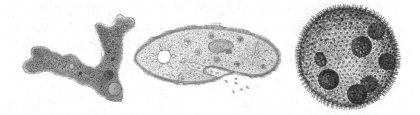

Fig 1   Left to right: amoeba, *Paramecium*, *Volvox*.

When so viewed, this particular amoeba appears as an irregularly shaped blob with several protrusions, like rudimentary tentacles, with a rather rounded shape. The outside of the creature is some sort of membrane, forming a flexible bag; the inside is a mixture of various granules, and a few holes, which flow with apparent purpose, like a thick jelly dotted with grains of sand that seem to know where they want to go. One rounded feature, dotted with even smaller particles, stands out: this is the nucleus. An amoeba can move and ingest food, and thanks to its nucleus it can even reproduce – its famous ability to 'multiply by division'. Under the right conditions the nucleus orchestrates a complex sequence of events that cause one amoeba to split into two. These in turn can grow, and divide again, so the amoeba's lineage can flourish.

One of my favourite cartoons shows the archetypal Noah's Ark, propped up by wooden scaffolding, the rain bucketing down. The last few pairs of animals are making their way up the gangplank into the ark, wet and miserable. Noah is grubbing around in the mud at the foot of the gangplank, desperately looking for something. Mrs Noah is leaning over the side of the ark, shouting: 'Noah! Forget the other amoeba!'

Van Leeuwenhoek also saw *Paramecium*, a slipper-shaped organism covered in tiny whip-like protrusions known as cilia (plural of cilium). These undergo wave-like motions and move the animal around. *Paramecium* also has a surrounding membrane. There is a mouth-like groove at one end and an anal pore at the other. It also has a relatively large nucleus, now called a macronucleus because genetically it resembles a large number of distinct nuclei that have merged into a single body.

A third common inhabitant of water-drops is a plant: *Volvox*. A mature *Volvox* is a colony of single-celled algae, and each cell propels itself with a flagellum, a tail-like object which appears to wiggle from side to side. These colonies, which can number up to 50,000 individuals, are contained in a larger (though still microscopic) sphere made from a gelatinous protein. They are bright green because they contain chlorophyll, the substance that gives plants their green colour and, more crucially, allows them to turn sunlight into chemical energy.

All this, and much more, inside one drop of water? It was scarcely credible. The luminaries of the Royal Society found it wildly unlikely, but after a further four years people started looking for themselves instead of denouncing the idea as absurd. Van Leeuwenhoek was vindicated, and soon he was elected a Fellow of the Society.

He made a number of fundamental discoveries using his microscopes, but ultimately his most important works were the microscopes themselves, because other people could use them to make their own discoveries. Van Leeuwenhoek manufactured more than 500 lenses and built 400 different microscopes. The best of the nine surviving models magnifies objects up to 275 times, and some of his models may have been capable of 500-fold magnification. This is five times more than a standard modern laboratory optical microscope. Of course, today's microscopes are manufactured with greater precision, and include all sorts of extras, and much higher magnification is available if you really need it. But you can do a lot of biology with one of Van Leeuwenhoek's microscopes.

Van Leeuwenhoek was a Calvinist, and considered his discoveries to be evidence of the hidden wonders of God's creation. On a scientific level he disproved the prevailing belief that microscopic organisms were 'spontaneously generated' – arose from non-living materials of their own accord – by showing that, just like larger living creatures, they reproduced. It is ironic that the telescope, with which Galileo revealed new things about the distant cosmos, raised so many religious hackles, but the microscope, which opened up entirely new visions of life on this planet, was accepted without a qualm.

It was not to last, of course. But the deep religious and emotional divisions that would be triggered by Darwin and his successors lay 200 years in the future.

. . . .

Microscopy really began to take off, and biology with it, when Robert Hooke joined the fray. Hooke was an English polymath and natural philosopher – the term used in those days for 'scientist' – and he took up where Van Leeuwenhoek left off. He was in many ways the true father of microscopy. He was into *everything*, and he possessed immense energy. When Hooke embarked on a new project, the sparks flew.

Hooke was responsible for one of the iconic biological drawings, one that made a very clear point about the complexities of minute organisms. In his lavishly illustrated *Micrographia* of 1665 he presented observations that he had made with both microscope and telescope. One of the engravings shows what a flea looks like through a modestly powered microscope (see Figure 2). All his contemporaries were familiar with fleas, indeed on intimate terms with them, but to most people these irritating little beasts were just dark specks that jumped a lot and sucked blood. Hooke revealed how complex a flea really is. It looks like a diminutive armoured machine. It has long legs, which allow it to jump, and the legs are hairy. Its mouth parts, which suck the blood, are surprisingly complicated. Clearly there is more to a flea than just being a nuisance.

Hooke was responsible for an even more iconic drawing, also in the *Micrographia*. It showed a thin slice through an everyday substance, cork (see Figure 3). Cork is the bark of a tree, and it is strong and light. These two properties stem from its microscopic structure: it consists of innumerable tiny chambers. Hooke called

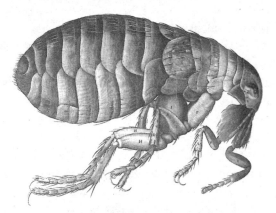

Fig 2   Hooke's drawing of a flea, from *Micrographia*.

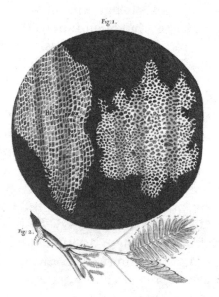

Fig 3   Hooke's drawing of cork, from *Micrographia*.

these chambers 'cells' because they reminded him of the rooms inhabited by monks. Cells are the basic building blocks of life.

Some organisms, such as the amoeba, are individual cells. Higher creatures, be they delphiniums, tigers or people, are huge assemblies of cells. At first it looked as though the crucial distinction between organisms was the number of cells they contained: one, or more. Single-celled organisms were simpler than many-celled ones. But when microscopists discovered how to see the different bits and pieces that made up a cell, they realised that there was a more fundamental difference. Some single-celled organisms – for instance, bacteria – were very different from other single-celled organisms, such as the amoeba. Most many-celled organisms belonged in the same category as some of the single-celled organisms, and the others were not so much organisms as colonies.

The fundamental distinction, in fact, is between prokaryotes and eukaryotes – two of the three 'domains' into which life is now classified. (The third domain is archaea, primitive single-celled creatures that used to be grouped with prokaryotes.) Eukaryote cells possess a nucleus; prokaryote cells don't. Bacteria are prokaryotes; amoebas and tigers are eukaryotes. Why so much fuss about the nucleus? Because it affects how the cell reproduces. All cells multiply by dividing: a single 'mother' cell splits, forming two 'daughter' cells. But prokaryotes do this in a much simpler manner than eukaryotes.

. . . .

When a cell divides, it splits into two pieces, each roughly half the size. Each piece is a new cell, a sort of copy of the original, and if necessary it can grow bigger. But reproduction must copy not just the form of the cell but the genetic information hidden inside it, because the genetics controls many of the processes that keep a cell alive. The genes are collected together in regions of the cell known as chromosomes, 'coloured bodies', a term that reflects their discovery when parts of the cell were selectively stained using dyes. When the cell divides, the chromosomes must somehow be copied, with one copy going to each daughter cell. This copying process is very different in prokaryotes and eukaryotes.

A prokaryote cell has a number of components (see Figure 4). Most of them are enclosed in an envelope – a bag that holds vital parts together. This has two layers: an outer cell wall and an inner membrane. The envelope is fairly rigid, so it helps the cell to maintain its shape. It is not totally impervious: some things are allowed in, some are allowed out. Its job is to control what goes each way. The outside is usually, but not always, decorated with structures that aid movement (flagella, plural of flagellum) and communication (pili, plural of pilus). A flagellum is a tail-like protuberance which can spin, propelling the cell through the

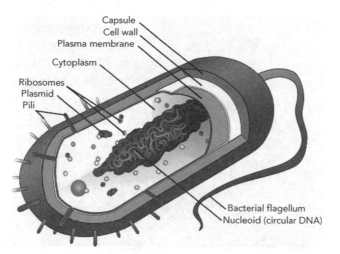

Fig 4   Prokaryote cell.

surrounding fluid medium. A pilus is a hair-like appendage, and cells can link their pili together, providing a communication channel between their interiors.

The region inside the cell envelope contains various special components, among them ribosomes, which make proteins, and the genetic material, which among other things specifies the structure of those proteins. The genetic material, which we now know is DNA, is almost always a long, closed loop, folded into a complicated tangle and attached to the membrane. There may also be free-floating loops of DNA, called plasmids. These permit 'bacterial sex', in which DNA is exchanged via the pili.

Eukaryote cells are more complex than prokaryotes, and usually larger, 10–15 times as wide and enclosing a thousand times the volume (see Figure 5). There is a cell membrane, but not always a cell wall. In place of flagella and pili there may be cilia, which wave from side to side, helping the cell to move. The most important difference is the genetic material. In a eukaryote cell, most of this is segregated inside a nucleus, which has its own membrane. It also consists of DNA, but now this molecule consists of long strands, not closed loops. The strands are organised by being wound round bobbin-like molecules called histones, and each strand forms a separate chromosome.

Eukaryote cells contain several other structures, known as organelles ('little organs'). Among them are ribosomes, which again make proteins, and mitochondria (plural of mitochondrion), which

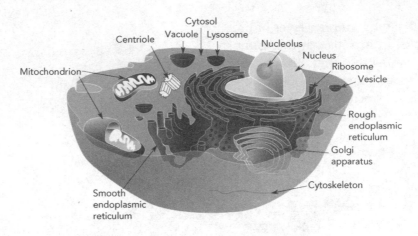

Fig 5 Eukaryote cell, with its organelles labelled.

manufacture a molecule called adenosine triphosphate (ATP), which in turn generates the cell's energy.

• • • • • • • • • • • • •

The ability of cells to move, when witnessed through a microscope, seems almost miraculous. They seem to know where they're going. However, we know enough about cellular movement to penetrate the apparent miracle and understand a little of what makes it tick. It depends on an organelle that controls the cell's shape; suitable changes in shape result in movement. The shape is maintained and changed using a kind of skeleton formed from long tubular molecules. These tubes can grow or come to pieces as required, and they are manufactured by another organelle, the centrosome.

The main agent of cellular movement is the cytoskeleton, a web of protein scaffolding inside the cell. It is built in part from microtubules – long, thin tubes made from a protein called tubulin. Tubulin occurs in two very similar but distinct forms, alpha- and beta-tubulin. The structure of a microtubule is like a chessboard rolled into a tube: the 'black' squares are alpha-tubulin and the white ones are beta-tubulin. Dynamically, this structure is unstable – it is like a cylindrical brick chimney in which successive rows of bricks fit precisely on top of one another instead of being staggered.

Why does nature make such an important item as a microtubule in such an unstable way? Because the 'cleavage lines', where the structure is weak, are useful. Microtubules can grow longer by adding another layer of protein bricks. But they can also shorten, splitting apart at the seams like a banana being peeled. Experiments show that they shorten about ten times as rapidly as they grow, and mathematical models of the forces that act between molecules and atoms support this observation. So the cell can go 'fishing' for interesting things using tubulin rods, pushing them out at random to see what they find, and collapsing them if they don't find anything. A cell moves by demolishing and rebuilding its own skeleton. It all boils down to the dynamics of a tiny molecular machine.

The construction and demolition of microtubules are controlled by chemical signals, many of which have an environmental origin. If the cell receives signals associated with food, it tears down its scaffolding on the side opposite the food, builds more of it on the side facing the food, and so inches its way foodwards.

Microtubules are produced by an organelle called the centrosome, first described by Theodor Boveri and Edouard van Beneden in 1887. When a cell divides, its chromosomes must replicate, and this process centres around a structure called the mitotic spindle. The chromosomes line up around the 'equator' of the mitotic spindle and subsequently migrate to its 'poles'. Through their microscopes, Boveri and van Beneden spotted a tiny dot at each pole of the mitotic spindle – a centrosome (see Figure 14, p. 87). A single cell has one centrosome, close to its nucleus. When the cell divides, the centrosome splits into two pieces which move apart. The mitotic spindle forms between them. Then the centrosomes pull the cell into two parts by extruding microtubules, which they use as fishing rods to grab chromosomes and pull them into the required positions using special chemical motors.

The centrosome consists of two identical molecular machines, the centrioles. Each centriole is a bundle of twenty-seven microtubules, arranged symmetrically in nine sets of three, glued together with a slight twist. Two of these devices, arranged at right angles to each other, are surrounded by a fuzzy cloud of 'pericentriolar material' from which sprout numerous tubulin fishing-rods. This elegant molecular machine also organises the production of new microtubules.

A combination of mathematical modelling and biochemistry has recently revealed yet another role for tubulin in a cell. Small molecules can diffuse through the cell unaided, but large, biologically important ones may not get to where they are needed if left to their own devices. A protein molecule known as kinesin 'walks' along the tubulin rods on little molecular legs,[1] carrying vital molecules across the cell. So a cell is not just a bag of chemicals – it is more like a highly automated factory.

. . . .

In prokaryotes and single-celled eukaryotes, the organism is a cell, so division of the cell constitutes reproduction of the organism. In many-celled higher life forms, including all the animals and plants familiar from everyday life and David Attenborough's television programmes, a lot more has to happen before the adult organism reproduces. In sexual organisms – the majority – male cells divide in a special way to produce sperm, and female cells similarly produce eggs. (I will describe this briefly later, when we have more background: both sperm and egg have half the normal amount of genetic material found in a cell.) These two types of specialist 'half-cell' are then brought together, the sperm fertilises the egg, and the two together form one conventional cell.

Once fertilised, the egg undergoes a complicated but organised pattern of *development*. In mammals, this takes it through a series of stages – embryo, fetus – leading to the point at which it emerges from its mother into the outside world. It then continues to develop through its juvenile stage until it becomes an adult. It is the same in birds and reptiles, except that for 'mother' you should read 'egg'. Other types of organism go through corresponding changes: for example, a frog develops through the tadpole stage and eventually turns into a miniature adult. Only at that point can the original adult organism be said to have reproduced.

Development is perhaps the most complex part of biology, because that is the stage at which we are forced to contemplate not just some isolated part of a living creature, but the whole creature. What we know about development is enormous, but what we don't know is far bigger. We know, in astonishing detail, how innumerable organisms develop – dogs, cats, dogfish, catfish, pigeons, spiders, marigolds, lizards, sea urchins, fruit flies, tiny nematode worms ... But we have far less understanding of the processes that control development.

The gist of it seems to be that the fertilised egg divides

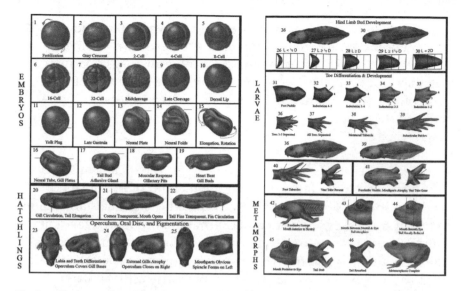

Fig 6   Early stages in development of the frog *Bufo valliceps*.

repeatedly, providing an ever-increasing number of cells, and that the organism's genetic material somehow orchestrates the patterns by which these cells grow, move, specialise to perform particular tasks and even die. Often the way nature makes a complicated structure involves using some cells as temporary scaffolding, and killing them off when they are no longer needed.

The early stages of development are similar in most of the higher animals; those of the frog are shown in Figure 6. The fertilised egg divides repeatedly without any change in the total size of all the cells, so the cells themselves become smaller and smaller. This process is known as cleavage, and it leads to a blastula, a hollow sphere of tiny cells. In many species this sphere is filled with fluid or yolk, but in mammals it contains another mass of cells, called the blastocyst.

Various layers of different cells appear; then a key step, known as gastrulation, occurs, and the entire ball of cells folds in on itself to create something more closely resembling a bag, or a tube with a hole at one end. (I'll describe a mathematical model of this process at the end of Chapter 13, which takes the cellular structure into account.) At this point the organism acquires an inside as well as an outside – so to speak. The internal organs can now form.

Next, we observe the beginnings of the nervous system. Two parallel ridges appear on the outside of the developing embryo, creating a groove between them; this is called the neural groove. It closes up into a tube, the neural tube, which later develops into the spinal cord and the nervous system. Another tube forms, which later becomes the digestive system, from mouth via stomach and intestines to anus. A rudimentary brain starts to appear ... and so on, and so on, and so on.

. . . .

Biochemistry alone cannot explain the complex changes of form that accompany development. We also have to take account of the physical properties of cells, such as how sticky they are, how they migrate from one region of the embryo to another, how new cells are born and how existing cells die. The appearance of stickiness in cells paved the way for the evolution of many-celled organisms; without it, they'd fall apart.

Development includes the deliberate destruction of cells that are employed as 'scaffolding' while some structure is forming, and are then destroyed once it has been made. This process is known as apoptosis, or programmed cell death. For example, an embryonic chicken's limbs develop from limb buds. At first the bud is just a single, featureless rounded shape, but after a time it splits into separate finger-like protrusions. This splitting is not just caused by separate protuberances growing: in the regions between them, cells die, much as a seamstress may make gloves by cutting away material between the fingers rather than sewing separate finger-shaped pieces of cloth together. Mathematical models have added to our understanding of limb growth, the shape of flies' wings, the tentacles of the tiny *Hydra*, and many other developmental puzzles.

Development is not just about molecular structure: its most important feature is shape. An organism cannot function effectively if its organs, limbs and body are the wrong shape. Biologists have learned a great deal about the changes that occur as an embryo develops. In insects, for example, large-scale structures such as legs and antennae develop from small regions of cells called imaginal discs. Experiments show that the growth and movement of these cells are controlled in part by specific genes, known as Hox genes. A mutation in one of these genes can, for instance, cause an antenna to form where a leg ought to be. Other genetic errors can create legs where there ought to be antennae.

Development involves an intricate interaction between genetics and physical processes of growth, movement and death. We are only just beginning to understand such processes, which pose a fascinating challenge for biologists, physicists, chemists and mathematicians.

3 Long List of Life

. . . .

This is a short chapter about a long list.

Early biologists studied life on a human level, considering an animal or a plant as a whole. They might dissect one, to see how it was arranged inside, but mostly they investigated life's diversity.

We all discover, as children, that a simple category like 'butterfly' fails to reflect the gaudy reality, with blue butterflies, red ones, brown ones, yellow ones, white ones, butterflies with spots, big butterflies, small butterflies, and so on. Even on a human scale, biology is huge. In order to grasp it, we have to cut it down into manageable chunks. It is too big and complicated for our village-level minds; we have to organise it. One way to handle the problem is to make a list. This was the second revolution.

. . . .

The extraordinary diversity of life on Earth is a relatively recent discovery, made only after intrepid scientists acquired the courage and the means to explore the furthest reaches of our planet and bring back specimens of what they had found.

I recently acquired a facsimile copy of the first edition of *Encyclopaedia Britannica*, which dates to 1771. About two-thirds of the way through Volume I (there are three altogether) is an entry for ARK. This includes a discussion of how many creatures Noah's Ark had to hold:

> The dimensions of the ark, as given by Moses, are 300 cubits in length, 50 in breadth, and 30 in height, which some have

thought too scanty, considering the number of things it was to contain; and hence an argument has been drawn against the authority of the relation. To solve this difficulty many of the ancient fathers, and their modern critics, have been put to very miserable shifts: But Buteo and Kircher have proved geometrically that, taking the common cubit of a foot and a half, the ark was abundantly sufficient for all the animals supposed to be lodged in it. Snellius computes the ark to have been above half an acre in area, and ... Dr Arbuthnot computes it to have been 81062 tuns.

The things contained in it were, besides eight persons of Noah's family, one pair of every species of unclean animals, and seven pair of every species of clean animals, with provisions for them all during the whole year. The former appears, at first view, almost infinite; but if we come to a calculation, the number of species of animals will be found to be much less than is generally imagined, not amounting to an hundred species of quadrupeds, nor two hundred of birds; out of which, in this case, are excepted such animals as can live in the water. Zoologists usually reckon but an hundred and seventy-two of the quadruped kind needed a place in the ark.

The article goes on to provide details of what sort of food would be needed for various animals, especially domestic ones, and to suggest a possible layout for the animal stalls and storage areas. It offers a startling insight into the thought processes of the period, and it makes some kind of sense given what was then known about the diversity of the species that live on our planet. But, without wishing to offend any sensibilities, the calculations were over-optimistic.

The Book of Genesis tells us that the Ark contained every species on the planet, though there is some ambiguity about creatures that live in water. However, an influx of enough fresh water to submerge the highest mountains would make the sea far less salty, hence unsuitable for sea creatures; conversely the extra salt in previously fresh water would kill off all the fresh water creatures. So *everything* would have to have gone into the Ark.

We now know that there are millions of species, not just a few hundred. Each would need its own special habitat, and food – which would often be other species. Even a common lion would need a five-month supply of gazelles. Then there's the leopard, the cheetah, the tiger, the jaguar, the serval, the lynx, the snow leopard, the fishing cat ... a total of 41 known species, and that's just cats.

I'm not trying to poke fun at the Noah tale, which is a charming moral fable derived from an earlier Babylonian flood story found in the *Epic of Gilgamesh*. My point is that less than 250 years ago, even the wisest scholars greatly underestimated the diversity of life on Earth, and let their personal beliefs blind them to the diversity in their own back garden, where a virtually endless parade of butterflies, moths and beetles – especially beetles – passed before their eyes every day.

. . . .

Some thinkers, however, were ahead of their time, and aware of the enormous diversity of nature. It was so diverse, in fact, that someone had to bring order to it if humans were ever to be able to understand it.

The first systematic approach to the classification of living organisms was the brainchild of a Swedish botanist, zoologist and doctor: Carl Linnaeus. To him we owe the standard system for naming organisms in terms of species, genus and more extensive groupings, using Latin (or Latinised) terms, a programme that he first put into practice in the 1740s – thirty years before that first edition of *Britannica*. In fact, the encyclopaedia has an extensive discussion of Linnaeus's classification of plants under BOTANY, and a shorter one for animals under NATURAL HISTORY. Linnaeus initially intended to include minerals, plants and animals in his classification, but it soon became clear that minerals were so different from living things that it was inappropriate to shoehorn them all into the same grand scheme. However, plants and animals are both forms of life, and although they have major differences they also have more in common than a quick glance might suggest. Many of the details of Linnaeus's scheme have changed considerably over the years, but the basic organisational principles remain the same.

The history of Linnaean classification, and the many changes that have occurred, is fascinating, but what matters for us is where it led. Today's taxonomists – biologists whose speciality is the classification of living organisms into species and related groupings – organise the living kingdom into an eight-tier hierarchy:

- life splits into three domains;
- each domain splits into kingdoms;
- each kingdom splits into phyla (plural of phylum);   Tun
- each phylum splits into classes;
- each class splits into orders;   orpuy
- each order splits into families;
- each family splits into genera (plural of genus);   pof
- each genus splits into species.

There are further divisions into subspecies and so on, but these are the eight main taxonomic ranks.

Looking at this list from the bottom up, species represent the different animals, birds, fish, plants, and so on. To a great extent they agree with our gut instinct that, say, all blue tits are basically the same type of bird, but thrushes are different. A few years ago some taxonomists compared the names used by natives of New Guinea for various birds to the names in the Linnaean classification, and both made exactly the same distinctions. The next level up, the genus, similarly corresponds to the view that blue tits and great tits and coal tits and so on are all variations on the theme of 'tit', whereas song thrushes and mistle thrushes are variations on the theme of 'thrush', but blue tits aren't. However, the genus on the whole makes finer distinctions than that: ducks, for example, fall into more than one genus. Families often reflect our instinctive opinions more closely.

More precisely, the blue tit is classified as in Table 1 (see over). This complete classification places the blue tit in a very specific relation to all other organisms – for example, the frog is also a chordate but not a bird, whereas the dandelion is a eukaryote but not an animal. (Eukaryotes have cells with nuclei; chordates develop a notochord, a precursor of the spinal column, as an embryo.) However, the full list is a bit of a mouthful, and for most purposes the final two groups suffice, the famous binomial (double-barrelled) classification, in which the blue tit is *Cyanistes caeruleus*, written like that in italics, with a capital letter for the genus but not for the species.[1] After the first mention the genus is usually abbreviated: *C. caeruleus*.

Table 1 Classification of the blue tit.

Domain	*Eukaryota*	Eukaryotes
Kingdom	*Animalia*	Animals
Phylum	*Chordata*	Chordates
Class	*Aves*	Birds
Order	*Passeriformes*	Perching or songbirds
Family	*Paridae*	Tits
Genus	*Cyanistes*	A subset of smaller tits
Species	*Caeruleus*	The blue tit

• • • • • • • • • • • •

Classification, however, is just the start of the complexity of biology – it is mere 'butterfly collecting' (which for lepidopterists it literally is). There is more to biology than just listing creatures and giving them fancy names. And the complexity of life is not just a matter of the quantity of different life forms, gigantic though that number is. Each individual organism, even the simplest, has enormous internal complexity. And when it comes to organisms interacting with each other in 'the environment' ... well, the magnitude of the task becomes almost overwhelming.

Nevertheless, classification is a sensible first step: it pins down the area of discourse and provides a basis for deeper comparisons and the search for general patterns. Many sciences could not have arisen without an initial stage of 'butterfly collecting'; a clear example is crystallography.

Taxonomists have so far listed just over one and a half million distinct species. They range in size from viruses to blue whales; they live in virtually every region of the planet, from boiling-hot vents in the ocean floor to clouds high in the stratosphere; they can be found in equatorial rainforests, deserts, rivers, lakes, seas, caves ... even miles underground in minute cracks in the rocks. About the only place where life has not yet turned up is in the molten magma of volcanoes – and given all the unlikely places where life has been found, in flat contradiction to what most scientists had previously thought was possible, it wouldn't be *too* surprising to find some exotic life form there as well. It would have to be a kind of life never before detected on Earth, and I doubt that anyone would lose their shirt betting against it.

Taxonomists currently recognise about 300,000 species of

plants, 30,000 fungi and other non-animals, and 1.25 million animals. Of these animals, 1.2 million are invertebrates – creatures lacking a backbone, such as snails and shrimps, of which some 400,000 are beetles. The geneticist and evolutionary biologist J.B.S. Haldane, asked by a lady what his studies had taught him about God, allegedly replied, 'That he has an inordinate fondness for beetles, madam.' Vertebrates account for a mere 60,000 species: 30,000 fish, 6,000 amphibians, 800 reptiles, 10,000 birds and 5,000-plus mammals. Among the mammals, about 630 species are primates, the order of animals that includes monkeys, lemurs, apes . . . and humans. In the last decade, 53 new species of primates have been discovered: 40 in Madagascar, two in Africa, three in Asia, and eight in Central and South America. Such discoveries are surprising in a world so thoroughly explored, but living creatures can be very elusive: they've evolved to be.

Out of all this enormous number of species, just one has developed reading, writing, religion, science, technology and language: *Homo sapiens*, human beings. Rudiments of most human attributes can be found in other creatures, and many of the more intelligent animals – such as chimpanzees and dolphins – are much smarter than we used to think even a few years ago. For that matter, so are crows.

How many species are there altogether? Estimates range from 2 million to 100 million, though a figure of 5 to 10 million is probable. A recent article plumped for 5.5 million, suggesting that previous estimates may have exaggerated the level of diversity.

Species are becoming extinct faster than they can be discovered. It is not entirely clear how we should define 'species'; in fact, it is not entirely clear that 'species' is a biologically meaningful concept at all. In my schooldays I was taught that there were two species of elephant, African and Indian. Today zoologists recognise five. In ten years' time . . . who knows?

. . . .

Linnaeus's classification scheme has brought a degree of order into the apparently chaotic world of life on Earth today. As an unexpected bonus, its hierarchical structure also hints at the evolutionary ancestry of today's organisms.

Nothing, however, is sacred in science, and a vocal minority of taxonomists feel that the world of living creatures is not as neat and tidy as Linnaeus's artificial scheme suggests. More than a dozen alternatives have been proposed, in which *Homo sapiens* becomes Homo-sapiens, homo.sapiens, homosapiens, sapiens1, sapiens0127654, and so on. The advantages claimed for such systems are that they reflect the complex reality of life, instead of shoehorning it into rigid, tidy categories.

Although these criticisms have some validity, Linnaeus's scheme – in its modern form – is convenient for the human mind, and has been in use for so long now that changing it would be extremely inconvenient. The widespread resistance to new, allegedly more rational systems is not just scientific conservatism: it is based on the realisation of how much effort would be needed to make the change. Many of the new schemes have flaws of their own, in any case. But in the long run, a scheme invented in the eighteenth century, when evolution, DNA and modern classification techniques did not exist, may well turn out not to be appropriate for the twenty-first century.

. . . .

Linnaeus's ideas made zoologists and botanists think more carefully about characters: the features that distinguish one species from another. Which characters are best suited for classifying organisms? Tigers and zebras are both striped, but that doesn't imply that they are closely related. In fact, tigers and zebras do not belong to the same genus, to the same family or even to the same order. Tigers are in the order *Carnivora* (carnivores), but zebras are in the order *Perissodactyla* (odd-toed hoofed animals). The two species come together only on the level of their class: both are mammals. So characters that strike the eye, like the tiger's stripes, are often less significant than subtler ones, such as how many toes the creature possesses.

The more widely a feature is shared, the higher the level of the corresponding taxonomic rank is likely to be, in the sense that classes are higher than orders and orders are higher than families. Higher ranks are more comprehensive. Many different animals produce milk and suckle their young. This is a key feature of all mammals, and because it is so widespread it takes precedence over more superficial characters such as coloration and markings. So what matters most about a tiger is that it is an animal, not a plant (kingdom); among animals it is a chordate (phylum); among chordates it is a mammal (class); then it is a carnivore (order), then a cat (family), then a big cat (genus). Only then, at the species level, do its iconic stripes enter the picture. Correspondingly, what matters most about a zebra is that it, too, is a mammal, but instead of being a carnivore it has hooves with an odd number of toes (order), is horse-like (family) and is *very* horse-like (genus). Stripes are shared by three distinct zebra species, and further characters are required to separate them.

Taxonomists quickly learned that the most important features for classification were seldom those that immediately attracted the attention of a human observer. Apparently minor features were particularly significant in flowering plants: a gigantic tree and a diminutive weed might be closely related, but two huge trees in the same forest might be totally different. What mattered most was often the tiny details of the reproductive organs of the plant – pistils, stamens, sepals and petals.

Initially, in his *Systema Sexuale*, Linnaeus grouped flowering plants according to how many of these various organs it had. He named the classes of plants *Monandria*, *Diandria*, *Triandria*, *Tetrandria*, and so on. He mainly did this for convenience: it was easy to count how many stamens or petals a flower had, and that made the system useful for identification. This classification was still popular in the mid-nineteenth century, for that purpose, but by then taxonomists had replaced it by a scheme that reflected the relationships among plants more faithfully. However, reproduction is a fundamental feature of plants, so the structure and number of reproductive organs are still important in the classification of plants.

Counting plant organs gave rise to one of the first extensive applications of mathematics to a problem in biology: striking patterns of numbers and shapes observed in the leaves and flowers of plants. The next chapter outlines the story, first using the kind of mathematics that was available in the nineteenth century and the early twentieth, then moving ahead to the modern era to see how the viewpoint has changed as new biological discoveries have motivated new questions for mathematicians to answer.

4   Florally Finding Fibonacci

· ·

The first two revolutions in biology set the pattern of subsequent research for over a century. The accepted way to advance biological knowledge was to seek out new species and fit them into the Linnaean scheme; to study them in detail, using a microscope where necessary, and to record and report what you found. This was the 'butterfly collecting' era of biology, when the aim was to catalogue life's diversity and celebrate its richness.

Amid the deluge of detail, a few general principles started to emerge, especially when it came to general relationships between organisms such as predator/prey, parasite/host, mimicry and symbiosis. These concepts helped to organise an ever-growing body of knowledge. But the predominant model was to collect, catalogue and observe. You weren't a biologist: you were a botanist (specialising in plants), a zoologist (animals), an entomologist (insects), a herpetologist (snakes), an ichthyologist (fishes), and so on.

The physical sciences followed a very different path. The same period, from Linnaeus in 1735 to Darwin in 1859, witnessed the explosive growth of physics, fuelled by the discovery of universal laws of nature, expressed in the arcane language of mathematics. But while physics was being unified by general mathematical principles, biology was being overwhelmed by a morass of individual examples. There were few general principles, hardly any laws to speak of and virtually no mathematics.

Even so, mathematics managed to squeeze itself into a few areas of biology, notably the strange numerology of the plant kingdom. A very specific sequence of numbers shows up repeatedly in association with plants, in several different contexts: the number of petals in a flower, the geometry of seed heads, the arrangement of leaves along a stem, the lumps on a cauliflower, and the way pineapples and pine cones fit together.

With the mathematical techniques available and the view of biology that prevailed around 1850, these numerical patterns could be – and were – described in considerable detail. Description, though, was as far as it went. Explaining *why* the patterns occurred was another matter entirely, beyond the scope of the science of that period. In this chapter, we will see how far Victorian-era science managed to get when grappling with the numerology and geometry of plants. Then, temporarily diverting from the historical story, we'll see how modern mathematics and a dash of chemistry have filled some of the gaps.

· · · · · · · · · · · ·

Marigolds typically have 13 petals. Asters have 21. Many daisies have 34 petals; if not, they usually have 55 or 89. Often, especially in cultivated plant varieties, the number of petals is twice as large, because plant breeders have learned how to double up the petals. On the whole, though, you seldom see a daisy with 37 petals, and if you see one with 33 then a petal has probably fallen off. Sunflowers, which also belong to the daisy family, usually have 55, 89 or 144 petals. Exceptions are usually the result of damage or disturbance while the young plant is maturing.

At first sight, there seems to be no particular reason why nature should favour these numbers – or any specific number, come to that. Petals are arranged like the spokes of a wheel around the central region of the flower. Just as a wheel can have any number of spokes, there seems to be no obvious restriction on how many petals can or cannot fit into the available space. This makes the limited list of numbers distinctly mysterious.

From today's gene-centred viewpoint, a plant could presumably have any reasonable number of petals, depending on the 'instructions' coded in its genes. So we would expect to find specific numbers for particular species, but not the same small list of rather strange numbers for many different species. However, this is what nature supplies, as Table 2 (see over) indicates. Other numbers are much rarer, though they do occur: for example, fuchsias have 4

Table 2 The number of petals on a flower.

No. of petals	Flowers
3	Iris, lily
5	Buttercup, columbine, larkspur, pink, wild rose
8	Coreopsis, delphinium
13	Cineraria, marigold, ragwort
21	Aster, black-eyed susan, chicory
34	Plantain, daisy, pyrethrum
55	Daisy, sunflower
89	Daisy, sunflower
144	Sunflower

petals. Exceptions like this often involve the numbers 4, 7, 11, 18 and 29, and I'll return to these later because they actually confirm more modern theories, rather than disproving them.

To compound the mystery, the same curious numbers show up elsewhere in the plant kingdom. A notable example is the arrangement of leaves on the stem of a plant, technically known as phyllotaxis. Some plants use very simple arrangements, with leaves arranged in pairs, one on either side. But many arrange their leaves in a helix, so that successive leaves along the stem are placed at a specific angle relative to the previous leaf. And those angles involve the same list of special numbers.

A typical case occurs for an angle of 135°, which is 3/8 of a full circle. If we say that the first leaf is at angle 0°, then the second will be at 135°, the third at 270°, and so on. The angles of successive leaves are the integer multiples of 135°. Subtracting 360° whenever the numbers become larger than a full turn, we obtain the sequence of angles (with the corresponding fractions of a full circle listed underneath):

0°	135°	270°	45°	180°	315°	90°	225°	0°
0	3/8	6/8	1/8	4/8	7/8	2/8	5/8	0

This pattern then repeats indefinitely. (I've left fractions like 6/8 as they are instead of reducing them to 3/4 to make the pattern clearer.) Figure 7 shows how these angles create a helical arrangement of leaves.

The same kind of behaviour occurs for many other plants, but several different fractions of a full turn can occur. However, other simple fractions, such as 2/7, are conspicuously absent. The ones

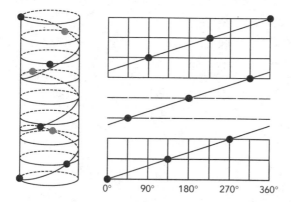

Fig 7 *Left*: The helical arrangement of successive leaves separated by 3/8 of a turn. *Right*: The same helix on the cylindrical surface of the plant stem rolled out flat. Note that 360° is equivalent to 0°, so the two ends 'wrap round' and join.

Table 3 **Fractions of a turn between successive leaves in different plants.**

Fraction of a turn between successive leaves	Plant
1/2	Grasses
1/3	Beech, hazel
2/5	Oak, apricot
3/8	Poplar, pear
5/13	Willow, almond

listed in Table 3 are closely related to the numbers observed in petals. In fact, each fraction in the list is formed from two of the numbers 1, 2, 3, 5, 8, 13. Apart from 1 and 2, these are all petal numbers. This is not a complete surprise, because petals are modified leaves, but it requires explanation.

Exactly the same numbers show up in several other features of plants, adding to the evidence that these patterns are not mere coincidence. Pineapples are easily recognised by the roughly hexagonal pattern on their surface. The hexagons are individual fruits, which coalesce as they grow. They fit neatly together, not into the standard honeycomb tiling, but into two interlocking families of helical spirals. One family winds anticlockwise, viewed from above, and contains 8 spirals; the other winds clockwise, and contains 13. It is also possible to see a third family of 5 spirals, winding clockwise at a shallower angle (Figure 8, see over). The

Fig 8   Three families of spirals in a pineapple.

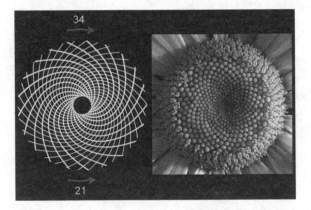

Fig 9   Two families of spirals in the head of a sunflower: 34 wind clockwise, 21 anticlockwise.

scales of pine cones form similar sets of spirals. So do the seeds in the head of a ripe sunflower (Figure 9), but now the spirals are not helical, but lie in a plane.

Clearly there is something about this list of numbers that makes them unusually suitable for plant structure. But why are these numbers so common, while others are much rarer?

. . . .

The answer began with rabbits.

In 1202 the Italian mathematician Leonardo of Pisa wrote a textbook on arithmetic. One of the exercises he set his readers was a problem about the progeny of a pair of rabbits. I'll discuss this problem and its answer in more detail in Chapter 16, when we take a look at the mathematics of population growth. Here we need to focus only on the resulting sequence of numbers, which lists how many pairs of rabbits there are in successive breeding seasons:

1, 1, 2, 3, 5, 8, 13, 21, 34, 55, 89, 144, 233, 377

and so on. The rule for forming these numbers (other than the first two 1's, which we take as a starting point) is that each is obtained by adding the previous two: 2=1+1, 3=1+2, 5=2+3, 8=3+5, 13=5+8, and so on. Leonardo later acquired the nickname Fibonacci, and ever since 1877, when the French mathematician and populariser Édouard Lucas wrote about this sequence, its members have been known as the Fibonacci numbers.

To paraphrase J.B.S. Haldane, the plant kingdom seems to have an inordinate fondness for Fibonacci numbers.

Leonardo's association of this sequence with rabbit progeny, though superficially biological, is completely irrelevant to petal numbers and phyllotaxis. Indeed, its assumptions are so artificial that it's not terribly relevant to rabbits either. But one mathematical feature of the answer is distinctly relevant: the fractions formed by successive Fibonacci numbers. If we write these fractions as decimals, something emerges:

1/1=1.000,   2/1=2.000,   3/2=1.500,   5/3=1.666,   8/5=1.600,
13/8=1.625,   21/13=1.615,   34/21=1.619,   55/34=1.617

As the numbers increase, the fraction gets closer and closer to a particular value, which to six decimal places is 1.618034. In fact, it is exactly equal to $(1+\sqrt{5})/2$, and is usually denoted by the symbol ϕ (the Greek letter phi). This number is irrational: that is, it cannot be represented by the ratio of two integers – no fraction can be exactly equal to it. In fact, it was one of the earliest irrational numbers discovered, after the square root of two, and it was known to the ancient Greeks in the context of the geometry of pentagons.

The fractions that appear in phyllotaxis are also ratios of Fibonacci numbers, but instead of using consecutive members of the sequence, the numerator (top) and denominator (bottom) are spaced two steps apart. Moreover, the larger number is the denominator, not the numerator. A typical case is 5/13, constructed from the portion of the Fibonacci sequence that reads 5, 8, 13. The first few fractions of this type are:

1/2=0.500, 1/3=0.333, 2/5=0.400, 3/8=0.375, 5/13=0.384,
8/21=0.380, 13/34=0.382, 21/55=0.381, 34/89=0.382

Again we see a pattern: the fractions get closer and closer to a particular number, this time 0.381966. This number is closely related to ϕ. In fact, some simple algebra shows that the exact value is $2-\phi$.

The special properties of the golden number ϕ provide an alternative interpretation. Suppose we divide a full circle (360°) into two arcs that are in golden ratio. That is, the angle determined by the larger arc is ϕ times the angle determined by the smaller arc. Then the smaller arc is $1/(1+\phi)$ times a full circle. Numerically, this expression is 0.381966: the number derived above. A bit more algebra confirms that this relationship is exact. Numerically, this angle is very close to 137.5°, and it is called the golden angle.

The upshot of all this is that we can interpret the fractions observed in phyllotaxis as approximations to the golden angle. That's all very well, but so far we have just replaced one mathematical puzzle by two others. The occurrence of Fibonacci numbers in flowers now seems to depend on a special angle and a sequence of related fractional approximations. Why this angle, and why these fractions?

The fractions are the easier part: they can be characterised mathematically as the *best* fractional approximations to the golden angle, for a given size of denominator. Not just in the sense that, say, 3/8 is a closer approximation than 2/8 or 4/8, but that if you look at fractions with larger denominators, the first time you get a closer approximation is when you reach 5/13. After that, the next improvement comes at 8/21, and so on. Classical mathematical concepts known as 'continued fractions' establish this relationship between the golden angle and the phyllotaxis fractions. So the key to the mystery is to understand why the golden angle appears. If we can do that, the role of the fractions should follow.

. . . .

The next step in solving the riddle of phyllotaxis was biological: specifically, taking a look at how the shoot of a growing plant changes on the cellular level. In 1868 the German botanist Wilhelm Hofmeister made extensive observations of this process, and laid the foundations for all subsequent work on the problem.

In the early stage of development, a plant appears to be little more than a tiny green shoot, with little structure. As it grows, small leaves begin to appear, and are 'left behind' as the shoot heads skyward. The main changes occur at the tip of the shoot. We can get a reasonable mental image of the growth pattern by thinking of a fountain, where water sprays upwards from the centre, moves outward radially and then starts to fall back towards the surface of the pond. Now suppose that the entire fountain heads skywards like a rocket, and the water that trails behind it 'freezes' once it drops below the level of the fountain's centre. Then you would see a growing column of frozen water, with a spurting fountain, climbing towards the heavens, perched precariously on top of it. All of the 'new' water would be produced by the fountain at the tip of the column, from which it would migrate radially until it reached the column's edge and froze.

A growing shoot is like that, but using new cells in place of droplets of water. For simplicity we can think of the shoot as a cylinder with a rounded top. Most new growth occurs near the tip of the shoot, close to the centre of the stem. New cells appear, through cell division, near the centre of the tip, and they migrate outwards towards the edge, where (at this level of description) they stop. In this way the growing tip pushes upwards, leaving a trail of new cells behind it, and the cylindrical column becomes taller without getting thicker.

As the plant matures, of course, the stem does get thicker, and many other changes occur, such as leaves getting bigger, buds appearing, and so on. But the explanation of phyllotaxis does not depend on these later changes: the basic pattern of leaf development – and much else – is determined by what happens at the growing tip.

The events unfolding there would be invisible without a microscope, because they involve small clumps of cells known as primordia (plural of primordium). Each clump will eventually become a leaf, so the positions of the leaves are set by the microscopic geometry of the primordia. Hofmeister discovered that the process begins with the appearance of two primordia, located together at the centre of the tip, on opposite sides. As these two start to migrate radially outwards, a third primordium appears near the centre, between them. There is insufficient room for this new primordium to grow to full size in that location, and the previous two push it away from the centre into an open space. The interplay of these forces causes the three primordia to arrange themselves so that the angles between the first and second, and the second and third, are close to the golden angle.

Before these three primordia have moved very far, a fourth appears near the centre of the tip. Because three times the golden angle is a little bit more than a full turn of 360°, this fourth primordium appears close to the first one and pushes it outwards. Then a fifth appears near the centre, and pushes against the second, as shown in Figure 10. The result of all this pushing and shoving, with new primordia popping into existence in the middle while the others move slowly outwards, is a beautiful geometric pattern. Successive primordia are spaced at multiples of the golden angle along a spiral. The shape of the spiral is determined by the rate at which the primordia move, and grow, so Hofmeister called it the generative spiral.

Shorn of the biological detail, we now have a mathematical description of the process that creates the geometrical and numerical patterns in phyllotaxis. Successive primordia lie on the generative spiral, each separated from its successor by the golden angle or a close approximation to it. As time passes, any given

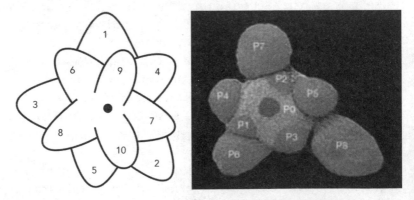

Fig 10   *Left:* Theory. Successive primordia (numbered in order of appearance) lie at a fixed angle to the previous one, and slowly migrate outwards. *Right:* Experiment. The growing tip of *Arabidopsis* (cress) as seen by an electron microscope, showing successive primordia (P8–P1 – the numbers are in reverse order compared to the previous picture). The next primordium will appear at P0. The region surrounding P0 is the source of new cells.

primordium migrates radially until it reaches the edge of the stem's rounded tip, at which point it stops moving. As the shoot grows, the result is a series of primordia, spaced at integer multiples of the golden angle along a helix that winds up the cylinder that represents the stem.

Victorian mathematicians, among them the mathematical physicist Peter Guthrie Tait, best known for his *Treatise on Natural Philosophy*, turned Hofmeister's ideas into a tidy mathematical description of phyllotaxis, based on diagrams like Figure 7 (see p. 41). But description was as far as Victorian mathematics got. At least one crucial question remained open: what is so special about the golden angle when it comes to plant growth? Yes, the golden angle comes from the golden number, but why is the golden number relevant?

In *On Growth and Form*, Thompson discusses one popular answer from the Victorian era: since the golden number is irrational, such an arrangement prevents any leaf from being exactly above another, allegedly allowing rain and sunlight to reach the leaves more effectively. He points out that the same argument works for any irrational number, and that the golden number can be obtained from many number sequences, not just the Fibonacci sequence. (He might also have pointed out that the approximations such as 5/13 that appear in real plants, which we are trying to explain, are rational.) He remarks, dismissively, that 'all such speculations as these hark back to a school of mystical idealism'.

In 1917, when Thompson's book first appeared, this must have seemed a shrewd comment. Golden numbers and Fibonacci sequences had a cultish appeal, and gave rise to an extensive literature that was long on speculation and short on fact. However, more recent work has established that the golden angle is a genuine feature of plant numerology, and so are its Fibonacci-fraction approximations. The reasons go beyond number mysticism by moving on from merely describing the geometry, and taking account of the dynamics of the growing plant. Progress on this problem has occurred in a series of steps, and we can take the story a little further here by temporarily following more modern developments, before returning to the historical order in the next chapter, with the third revolution in biology.

• • • • • • • • • • • •

The special features of the golden angle are easier to appreciate if we turn to a related feature of plants: not the positioning of primordia as the young shoot first grows, but the positioning of seeds in the flower of a mature plant. Fibonacci numbers are to be found here as well. So is Hofmeister's generative spiral, because the locations of the seeds are determined by patterns of primordia in the young shoot. These patterns are not activated until the plant matures, but they are created by the same mechanism that generates leaves. So again, the main issue is the geometry of primordia, which generate not only the leaves of the plant but many of the other interesting organs, including the petals, and the seeds in the seed-head.

We've already seen in Figure 9 (see p. 42) that the seeds in a sunflower head are packed together in such a way that the human eye is immediately attracted to two interpenetrating families of spirals, one running clockwise, the other anticlockwise. If you count how many spirals there are in each family, you typically find two consecutive Fibonacci numbers. By the way, these spirals are not the generative spiral, which is more loosely wound and not apparent to the eye – unless you join primordia in the order in which they form.

Petals form at the outer end of one family of spirals, again cryptically determined by the original pattern of primordia, which are specialised to form petals, not seeds. So a Fibonacci number of spirals implies a Fibonacci number of petals. In short, it is enough to explain why Fibonacci numbers turn up in the spirals.

In 1979, Helmut Vogel of the Technical University of Munich considered a simple mathematical representation of the geometry of sunflower seeds, and used it to explain why the golden angle is especially suited to such arrangements.[1] In his model, the nth primordium is placed at an angle equal to n times $137.5°$, and its distance from the centre is proportional to the square root of n. These two numbers determine its location, and Hofmeister's generative spiral is revealed as a so-called Fermat spiral, which becomes more tightly wound as it moves outwards from the centre.

Using this formula, Vogel worked out what would happen to the seed head if the same generative spiral were employed, but the golden angle of $137.5°$ were changed, ever so slightly. The result, shown in Figure 11, is striking. Only the golden angle leads to seeds

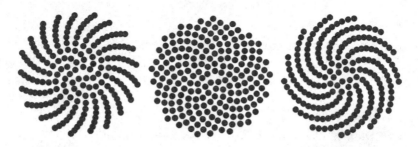

Fig 11   Fermat spiral patterns. *Left*: Spacing 137°, just less than the golden angle. *Middle*: Spacing 137.5°, the golden angle. *Right*: Spacing 138°, just greater than the golden angle.

that are packed closely together, with no gaps or overlaps. Even a change in the angle of one-tenth of a degree causes the pattern to break up into a single family of spirals, with gaps between the seeds.

Vogel's model explains why the golden angle is special, and supports the view that it plays an active role in phyllotaxis, and is not some numerical coincidence viewed through mystical eyes. However, the full explanation lies deeper. It turns out that – just as Hofmeister had said – the dynamics of the growing plant causes primordia to be *pushed* into golden-angle relationships. As the cells grow and move, they create forces that affect neighbouring cells.

. . . .

Forces are an essential ingredient of mechanics, which is the mathematical physics of moving objects. Mechanics was born in experiments that Galileo made around 1600, when he rolled balls down a slope to investigate the effects of gravity. It became a recognised branch of mathematics in 1687, when Newton published his epic *Philosophiae Naturalis Principia Mathematica* ('Mathematical Principles of Natural Philosophy') in which he related the motion of a body to the forces acting on it. It has since become a cornerstone of science.

Once mechanics enters the picture, attention switches from merely describing natural phenomena to investigating the mechanisms that cause them. The golden angle and its Fibonacci fractions cease to be numerical curiosities, and their presence is now explained by the interplay of forces in the growing stem.

Contrary to Thompson's scepticism, the golden number and the mathematics associated with it do, in fact, play key roles.

In 1992 the French mathematical physicists Stéphane Douady and Yves Couder investigated the mechanics of systems in which point-like objects representing primordia are repeatedly introduced near the centre of a circular disc at equally spaced instants of time, and then made to move outwards along a radius of the circle.[2] They assumed that these objects repel each other, much as two north poles of magnets do. Then they worked out what would happen in two ways: by experiment, and by computer simulation.

In their experiments, successive primordia were represented by droplets of magnetic fluid, under the action of a magnetic field that caused them to repel each other, so that they organised one another into patterns while migrating in a radial direction. The result depended on the strength of the magnetic field and the intervals between droplets, but in the most common pattern that developed the droplets spontaneously packed themselves into families of spirals just like those in the sunflower head, complete with golden angle and Fibonacci numerology.

Computer simulations revealed more detail, showing that the system of droplets naturally homes in on Fibonacci-fraction approximations to the golden angle. The fraction that emerges depends on the rate at which new droplets are added. The precise relation is captured by a so-called bifurcation diagram (Figure 12). This shows how the numbers in the spirals, and the associated

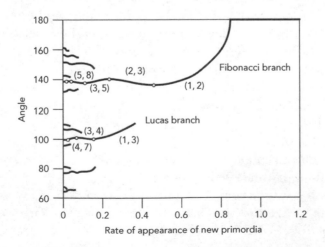

Fig 12   Bifurcation diagram for phyllotaxis, after Douady and Couder.

angle between successive primordia, relate to this rate. The main features of the bifurcation diagram are two curves, called the Fibonacci branch and the Lucas branch. There are other branches – in principle infinitely many – but these are very short, so they rarely occur in practice.

To interpret the curves, we start at the right hand end and move to the left. When the rate of droplet formation is bigger than 0.8, the angle is 180°, so successive droplets line up on either side of the centre. As the rate slows down, the angle changes continuously, following the curve that bends from the top of the picture down towards the left, and the angle changes to about 140°. This curve, the Fibonacci branch, corresponds to the most likely behaviour. It wiggles up and down, and at each peak or trough (shown by a white dot) the numbers of spirals in the two families move one step up the sequence. If we write, say (3, 5) to mean 3 families in one direction and 5 in the other, then these pairs of numbers change from (1, 2) to (2, 3), then to (3, 5), then to (5, 8), and so on, as the curve runs from right to left. The angle converges towards the golden angle of 137.5°, as expected.

Now the remaining pieces of the mathematical puzzle were falling into place: why the golden angle is the organising principle behind the geometry, *and* why what we actually observe are Fibonacci-fraction approximations. The main missing ingredient was a rigorous mathematical proof that the features seen in the simulations are what they appear to be. This was provided in 1991 by Leonid Levitov, a condensed matter physicist now at the Massachusetts Institute of Technology. In 1995 Martin Kunz, a physicist at the University of Lausanne in Switzerland, added many further details.[3]

These results also explained a puzzling feature of the problem which I have glossed over so far: the occasional appearance of non-Fibonacci numbers, such as the four petals of the fuchsia. These exceptions also appear in Douady and Couder's bifurcation diagram, but on the Lucas branch – so they have the same explanation as the Fibonacci numbers. Now the numbers of spirals come from a sequence very like the Fibonacci sequence, called the Lucas numbers:

1, 3, 4, 7, 11, 18, 29, 47, 76, 123

and so on. The rule for forming the numbers, after the first two, is the same as before: each number is the sum of the previous two, but here the first two numbers are 1 and 3, not 1 and 1. The ratios of successive Lucas numbers also approach the golden number. The pairs of numbers of spirals now become (1, 3), then (3, 4), then (4, 7), then (7, 11), and so on. The angle converges to 99.5°.

The other, shorter, branches in the bifurcation diagram correspond to patterns that follow sequences like the Fibonacci and Lucas sequences, but starting with different numbers. They correspond to angles such as 77.9° and 151.1°, and are hardly ever seen in plants.

The four petals of the fuchsia are not the only occurrence of Lucas numbers in plants. Some cacti exhibit a pattern of 4 spirals in one direction and 7 in the other, or 11 in one direction and 18 in the other. A species of echinocactus has 29 ribs.[4] Sets of 47 and 76 spirals have been found in sunflowers.[5]

• • • • • • • • • • • •

Cacti lead naturally to an extension of the mathematical analysis of plant structure. In the models I have just described, primordia are represented as discrete point-like objects, and the forces act on the associated points. In a more realistic 'continuum' model, the forces would be distributed over the entire surface of the growing stem, and the primordia would develop as a consequence of those forces, much as a sheet of metal will buckle when its edges are compressed. The techniques required here come from a major branch of applied mathematics: elasticity theory. This studies how shapes that are able to bend or compress behave when they are subjected to external forces; it is widely used by engineers when designing buildings, bridges and other large structures.

If you distort an elastic object, you have to do work. Think of squeezing a rubber ball, for instance. The work you do in squeezing the ball is stored in the material as a form of energy, known as elastic energy. A central principle in elasticity theory is that systems behave so as to minimise their elastic energy. In 2004, Patrick Shipman and Alan Newell, mathematicians at the University of Arizona in Tucson, applied elasticity theory to continuum models of a growing plant shoot, with particular emphasis on cacti – which are widespread in Arizona.[6] They modelled the formation of primordia as a kind of buckling of the surface of the tip of the growing shoot, and showed that minimum-energy configurations take the form of superimposed patterns of parallel waves.

These patterns are governed by two factors: the wave number, which is related to wavelength, and the direction in which the waves point. The Fibonacci numerology, in this approach, arises because the most important patterns involve the interaction of three such waves, and in the relevant states, the wave number for the third wave must be the sum of the other two wave numbers. The spirals on the pineapple in Figure 8 (see p. 42) show three systems of roughly parallel lines of hexagons: it's basically the same idea. So this model traces the arithmetic of Fibonacci numbers directly to the arithmetic of wave patterns. Not only the numbers, but even the mathematical rule for their formation, correspond directly to the underlying mechanics of the buckling tip.

Any botanist will tell you that plant tips don't really buckle: they grow. So although the elasticity model captured some of the main features of plant growth, it was still missing some crucial ingredient. The forces that act on primordia explain their geometry, but they don't explain how new primordia are produced, and why they appear at the places where elastic buckling says they should.

The answer to this question required not mathematics, but biochemistry. The formation of primordia is driven by a hormone, called auxin. Newell and colleagues have shown that similar wave patterns arise in the auxin distribution.[7] So the story as it is now understood involves an interplay among the biochemistry of the growing plant, the mechanical forces between cells, and the plant's geometry. Auxin stimulates the growth of new primordia. Primordia exert forces on one another and, in combination with the growth of the plant, these forces create the geometry. The geometry may also affect the plant's biochemistry, for example by triggering the production of extra auxin in specific places. So there is a complex set of feedback loops between biochemistry and mechanics, mechanics and geometry, and geometry and biochemistry – and all of these ingredients are required. Current mathematical theories therefore have to take into account many features of the biology and physics of the growing plant that were undreamt of in Victorian times.

• • • • • • • • • • • •

As a result of all this activity, we now know that D'Arcy Thompson's scepticism was unfounded. The role of the golden number in phyllotaxis is genuine and informative. The associated mathematics provides a convincing explanation – indeed, several complementary ones – of Fibonacci numerology. It also explains why some plants do *not* have Fibonacci numbers of petals, by predicting the rarer occurrence of Lucas numbers. These account for most of the exceptions seen in nature. So the Victorian work on the golden angle, as a description of phyllotaxis, has motivated deeper mathematical theories that are more broadly applicable. What seemed to be exceptions, a century ago, actually confirm the deeper mathematical theory that underpins the Victorian one.

However, many plants do not fit into even this more general description of phyllotaxis. Some even seem to produce leaves and other organs pretty much at random. So the story is still incomplete.

A word of warning is also in order. Thompson had a reason for his doubts, and it has not entirely gone away. Humanity's fascination with the golden number has often led to exaggerated claims for its importance, usually in contexts that are mathematically vague. Entire books have been written about the golden number in nature and art, finding it in the spirals in goats' horns and in the proportions of the Great Pyramid and the Parthenon. It is often stated that the most aesthetically pleasing shape for a rectangle occurs when its sides are in the golden ratio.

There seems to be very little basis for this claim: it is a mathematical urban myth. Many of these supposed occurrences of golden numerology are probably spurious or accidental. Some methods of statistical analysis can concentrate data around the golden number, exaggerating its significance. Any measurement close to 1.6 can be attributed to the golden number, but the relationship is likely to be coincidental, unless – as is now known to be the case for phyllotaxis – the phenomenon concerned arises from a deeper model in which the golden number turns up for solid structural reasons.

It is also often stated that the nautilus shell, which forms a beautiful spiral, is an example of the golden number in nature. This is a misunderstanding. The shape of the shell is impressively similar to a logarithmic spiral, in which successive turns of the spiral are magnified by a fixed amount (call it the growth rate). Moreover, there is an elegant logarithmic spiral whose growth rate is related to the golden number. However, there are many logarithmic spirals, with many different growth rates, and the nautilus has a different growth rate from the golden number spiral. So there is no meaningful relation between the golden number and the nautilus.

It would be truer to say that phyllotaxis is virtually the *only* context – aside from laboratory physics – in which the golden number can confidently be associated with the natural world. And even there, the connection is not universal. But we should not expect connections between mathematics and biology to be universally valid, subject to absolutely no exceptions. Biological systems are versatile and adaptable. Mathematical models will apply within some range of validity, but it's not sensible to expect them to apply everywhere.

Our excursion into Victorian and early-twentieth-century mathematical biology, with extra insight from its modern sequels, is now complete. We have seen what could be achieved on the basis of the first two biological revolutions. Now we return to the remaining three revolutions, which will set the scene for the dominant theme thereafter: mathematics.

5   The Origin of Species

· ·

Revolution number three got off to a bad start.

The year was 1858; the date, 1 July. The Linnaean Society, then and now the world's oldest society for natural history and taxonomy, had been in existence for seventy years. It was the final meeting of the session, and the members had their minds on summer holidays and outside activities. The president, Thomas Bell, was delivering his annual review of the highlights of the society's scientific activities. But it had been a bad year, in his opinion, and there hadn't been any highlights. 'The year which has passed,' said Bell, 'has not, indeed, been marked by any of those striking discoveries which at once revolutionize, so to speak, the department of science on which they bear'.[1]

At the time, no one objected to his summing up. Even the two scientific papers that had been squeezed into the programme of the meeting at the last minute had made no impact; when the members departed for their homes, no one seems to have been terribly impressed by them. As was the practice at the time, these papers were read out loud to the Society on behalf of the authors. They were on very similar topics, and their titles were 'On the tendency of species to form varieties' and 'On the perpetuation of varieties and species by natural means of selection'. Their authors, respectively, were Charles Darwin and Alfred Russel Wallace.

The two papers, deliberately presented simultaneously to avoid any priority disputes, announced the theory of evolution by means of natural selection.

. . . .

Long before Darwin was born, biologists were trying to understand how the planet's innumerable species had come to be. Almost everyone considered them to have been divinely created, the default social assumption of the time. But that is to answer every question with the same facile formula. Where do dogs come from? God created them. Where do dragons come from? God created them. Well, no, he didn't, but you wouldn't be able to deduce that from the answer.

In his *Physicae Auscultationes* ('Lectures on Nature'), the Greek philosopher Aristotle objected to explanations of nature that invoke purpose, such as 'rain falls in order to make corn grow'. He argued that if this were the case, then rain would also exist in order to spoil the famer's corn if he threshed it outdoors. Continuing this line of thought, he asked why animals' anatomical features are so obviously related to one another's. His answer is surprisingly modern: if anything didn't work in reasonable concert with the rest of the body, it would have been impossible for the animal to function, so neither the animal nor that combination of features would survive.

By the later eighteenth century, some scientists were beginning to think that over long periods of time, organisms could change. Among them was Darwin's grandfather, Erasmus. Professionally he was a physician, but he had the broad interests of a polymath, including natural history, physiology, abolition of the slave trade and inventing things. He was a founder member of the Lunar Society, a scientific society which met in Birmingham once a month on the night of the full moon, to make it easier for members to find their way home in the dark. His main claim to biological fame is the *Zoonomia* of 1794–6, in which he asked whether it would be too bold to imagine that

> in the great length of time since the earth began to exist, perhaps millions of ages before the commencement of the history of mankind ... all warm-blooded animals have arisen from one living filament, which the great First Cause endued with animality, with the power of acquiring new parts, attended with new propensities, directed by irritations, sensations, volitions and associations, and thus possessing the faculty of continuing to improve by its own inherent activity, and of

    delivering down these improvements by generation to its
    posterity.

Erasmus (let me call him that to distinguish him from his
grandson) was convinced that species could 'transmute' – that is,
change spontaneously – and that the process began with a single
primitive organism. Biologists now call this idea 'universal common
ancestry'. But Erasmus offered no specific mechanism that could
produce such changes.

    Darwin doesn't mention his grandfather's work in the *Origin*,
possibly because he found *Zoonomia* too eccentric, but probably
because he did not consider it relevant. (We know he had read it,
because he wrote the title on the opening page of his 'B Notebook',
his first recorded step towards the *Origin*.) Unlike his grandfather,
Darwin wanted to know *how* species changed. Erasmus seems to
have thought that animals could acquire new abilities, and that
these would automatically be passed on to their descendants. This
belief in the 'inheritance of acquired characters' was soon to be
advocated more explicitly by a better-known figure, whom history
has treated somewhat unfairly.

. . . .

Jean-Baptiste Lamarck, the eleventh child of an upper-class family
down on its luck, trained as a Jesuit but abandoned his studies to
join the French army, then at war with Prussia. When illness forced
him to retire from the military, he tried his hand at medicine and
banking, before settling on botany and becoming keeper of the
royal herbarium to King Louis XVI in 1788. He retained the
position, but not the royal connection, through the climax of the
French Revolution, whereas Louis's head failed to retain its
connection to his body. Lamarck then became curator and professor
of invertebrate zoology at the National Museum of Natural History.

    His most significant publications include the 1809 *Philosophie
Zoologique* ('Zoological Philosophy') and the seven-volume *Histoire
Naturelle des Animaux sans Vertèbres* ('Natural History of
Invertebrates') published between 1815 and 1822. In these books,
and elsewhere, he developed and elaborated a novel idea: animals
can change from generation to generation in response to their
environment. To Lamarck, moles were not blind because they had
been created that way, but because they lived underground and so

did not need the sense of sight. Their ancestors had once been sighted, but the ability to see had been lost because it was not needed. He tried to find a credible mechanism for such changes, among them being two 'forces': the tendency of living creatures to become more complex, and their tendency to adapt to their surroundings. He thought that living creatures ascended a ladder of progress, propelled by some inherent force that created ever-increasing order.

Today, Lamarck's name is most often associated with a discredited view of evolution, the 'inheritance of acquired characters': if some organism happens to develop a useful feature, such as a longer neck or stronger muscles, then this feature can and will be inherited by its descendants. Thus a blacksmith whose trade causes him to have very strong arms will have sons with strong arms – which was often true, because sons went into their fathers' trade. However, Lamarck did not believe that evolving organisms changed in a purposeful way, and he did not think that every acquired character would be passed on to future generations. He believed that all changes in organisms had a purely physical origin.

Lamarck distilled his view of adaptation into two laws:

1. If animals use an organ more often, that organ will become stronger and larger. Conversely, any organ that is not used will weaken and eventually disappear.

2. Any such improvement or loss, if it is related to the animals' long-term environment, will be passed on to future generations.

The second law is where the notion that Lamarck believed in the inheritance of acquired characters comes from. He did, but only for certain types of character. Darwin pointed to Lamarck's emphasis on use and disuse, and interpreted those aspects of Lamarck's work as a form of natural selection. He praised Lamarck for drawing attention to 'the probability of all change in the organic world being the result of law, not miraculous interposition'. In Darwin's view, Lamarck came close to a scientifically acceptable mechanism for evolution, but fell short.

. . . .

As a young man, Darwin was interested in geology, having been enormously impressed by the concept of deep time – the idea that the Earth is enormously old, whose significance was emphasised by Charles Lyell. None of this greatly impressed his father, who wanted Darwin to be a doctor and take his rightful place in Victorian society. But Darwin's stint as a medical student at the University of Edinburgh didn't work out, so his father decided that the son would do better to settle down and become a country vicar. That would leave him ample spare time to pursue his geological fancies. So in 1828 Darwin entered the University of Cambridge to study theology.

Unfortunately for his father's careful plans, Darwin was promptly bitten by a bug. Ironically, it came about because of a country vicar, William Kirby, who collaborated with the businessman William Spence on a four-volume treatise, *An Introduction to Entomology*. The book sparked a national craze for collecting beetles, and Darwin joined in with a passion, hoping to find a new species. He failed, but did find a rare German one. He also developed a second passion, a young woman named Fanny Owen, who ditched him as soon as she discovered he was more interested in beetles.

Neither interest did much for his examination preparations, and he found himself facing a backlog of two years' work with only two months to do it. One of the important course books was *Evidences of Christianity* by the Reverend William Paley. Darwin was captivated by its logic and its leftish politics. He scraped a pass and moved on to his final year. Now he had to read another book by Paley, *Principles of Moral and Political Philosophy* – not because of its orthodoxy, but because students had to learn how to argue *against* the book's assertions, such as the irrelevance of an established Church to Christianity.

Darwin decided to read around the topic, and chanced upon Paley's *Natural Theology*, which made the case for the divine creation of living creatures. He was impressed by the book's clarity, but he was also aware that many leading scientists and philosophers found it naive, and this led him to investigate the process that led to scientific laws, reading Sir John Herschel's *Preliminary Discourse on the Study of Natural Philosophy*. For light reading he scanned the 3,754 pages of Alexander von Humboldt's *Personal Narrative*, about the exploration of South America. From Herschel he learned how to do science; from Humboldt he learned where to do it. He promptly vowed that he would visit the volcanic Canary Islands to see the famed Great Dragon Tree.

This plan collapsed when his friend Marmaduke Ramsay, who was going to go with him, died unexpectedly. While Darwin was trying to work out what to do next, he was offered the post of gentleman companion to a naval officer, Robert FitzRoy. FitzRoy had been charged with carrying out a chronometric survey of the coast of South America – that is, a survey using a marine chronometer (essentially, a very accurate watch) to determine longitude. The ship was to be the *Beagle*, and FitzRoy was worried because the previous captain had shot himself. Worse, one of FitzRoy's uncles had slit his own throat when depressed. So FitzRoy determined to stave off suicide by taking along someone capable of intellectual discussion.

This suited Darwin down to the ground, but his father refused permission, until he received a letter from Darwin's uncle Josiah, saying that the trip would be the making of the young man. So off Darwin went, on what eventually became a five-year voyage round the world. First landfall was St Jago, a rugged volcanic outcrop of the Cape Verde Islands, with impressive volcanoes where Darwin could pursue his geological interests, and fertile valleys where he could do natural history. He found flatworms in Brazil, fossils in Argentina and naked savages in Tierra del Fuego. Similarities between shells on the beach and fossils high in the Chilean Cordillera convinced him that the Andes must have been pushed up high above sea level by vast geological forces. But the climax of the voyage, scientifically speaking, came when the *Beagle* arrived at the volcanic Galápagos Islands, which basically were a nest of volcanoes.

Darwin's stay there was brief, but it allowed him to collect specimens from what he quickly realised was very newly formed land. Many of its native creatures were bizarre: a species of penguin living on the equator; the only known marine iguanas, which foraged for algae beneath the ocean's turbulent waves; the only known species of cormorant that could not fly; and giant tortoises up to two metres in circumference. He was amazed by the spectacular blue-footed boobies, which dived into the sea from a great height, like avian arrows in search of fishy targets, and he was intrigued by the mockingbirds, which differed subtly from one island to the next. Looking more closely, he discovered that they formed at least three distinct species: one on Charles Island (now Santa Maria), another on Chatham Island (San Cristóbal), and a third on James Island (San Salvador).

He collected a few dull-looking finches and warblers, but found them rather boring.

The voyage continued to Tahiti, New Zealand and Australia. Five years and three days after the *Beagle* departed from Plymouth, a travel-weary Darwin arrived home. When he walked in through the door, he found his father eating breakfast, who greeted him without enthusiasm, remarking, 'Why, the shape of his head is quite altered.'

. . . .

Only after his return from the *Beagle* voyage did Darwin start to think seriously about what he had seen.

Near the end of the voyage, in Australia, he had made one major discovery: the origin of coral reefs. Lyell had suggested that reefs must be built on top of submerged volcanoes, but Darwin came up with a different idea: coral reefs start out in regions where the seas are shallow, but then the sea floor slowly falls. The corals grow faster than the seabed can drop, so the living tip of the reef remains near the surface. On the basis of this, and his Andes observations, he was made a Fellow of the Royal Geological Society. His scientific reputation was made not by evolution, but by geology.

However, Darwin's training as a geologist made him aware of some puzzling aspects of what he had observed. Lyell, who was a firm believer in divine creation, explained the diversity of living creatures, and their adaptation to their environments, in terms of local geological conditions. Darwin was sceptical. The Galápagos finches, which he had dismissed as uninteresting, were coming back to haunt him. He had misunderstood them. In fact, he had misunderstood them so badly that he hadn't even realised they were all finches – he thought some were wrens, and others blackbirds. On his return to England he immediately gave all the relevant specimens to John Gould, a finch expert at the Zoological Society. It took Gould only ten days to convince himself that they were all finches, very closely related, but constituting an

astonishing twelve distinct species (now considered to be thirteen).
Why so many species on such a small group of tiny islands?

Initially Darwin hadn't been interested in the question, but now
he began to take it seriously – and it bothered him. For once, his
skills as an observer had deserted him, perhaps for lack of time. He
hadn't recorded which specimens came from which island. And
he'd assumed that the finches formed huge flocks, which all fed on
the same things. But when he examined their beaks more closely,
he realised this must be wrong. There were many different forms of
beak, suited to different foods.

The Darwin family were Unitarians, which predisposed Darwin
to the belief that God worked only on vast scales of space and time.
In support of this view he wrote[2]

> It is derogatory that the Creator of countless systems of worlds
> should have created each of the myriads of creeping parasites
> and slimy worms which have swarmed each day of life on land
> and water on this one globe. We cease being astonished,
> however much we may deplore, that a group of animals should
> have been directly created to lay their eggs in bowels and flesh
> of others – that some organisms should delight in cruelty . . .
> From death, famine, rapine, and the concealed war of nature we
> can see that the highest good, which we can conceive, the
> creation of the higher animals has directly come.

Mainstream Victorian theologians were no longer in favour of
Paley's arguments, for similar reasons. If God continually
intervened in His creation, that seemed to suggest that He had
bungled it. Why else keep tinkering? The theist view was being
replaced by a deist one: yes, the Creator set up the universe, along
with its laws, but then He stood back and left it running, to work
out its own destiny according to those laws. And one consequence
of those laws seemed to be that species could change. Darwin had
been keeping a notebook, his Red Notebook. Now he started a new,
secret one, the B Notebook, on the transmutation of species. Slowly
he assembled a long list of puzzles that made much more sense if
species could change. But even if they did change, he still didn't
know how.

Back in England, Darwin wrote a series of books – two on the
Beagle voyage, one on coral reefs, one on the geology of South
America and a massive series of four huge tomes on barnacles. (He
had been advised to become an expert on some limited range of

organisms to cement his reputation as a naturalist, and settled on barnacles.) The barnacles added further weight to the argument against special creation – there were hundreds of different species, most of them very similar to many of the others, all minor variations on the same underlying theme. God's inordinate fondness for beetles now seemed to extend to an equally inordinate fondness for barnacles. Creation of each species, one by one, seemed absurd. So much neater to create just one, and then let it ... change.

. . . .

As Darwin's ideas on the transmutation of species began to crystallise, he realised that a simple mechanism might explain how it happened. The idea was triggered when he read Thomas Malthus's 1826 *Essay on the Principle of Population*. Malthus's argument relied on some simple mathematics. He asserted that populations of living creatures, if their growth is not restrained by lack of food or predation, grow 'geometrically': the population size at successive instants of time is multiplied by the same fixed amount. For example, if every pair of finches produces four surviving adults, then the finch population repeatedly doubles. The numbers grow very rapidly – the modern term is 'exponentially':

2, 4, 8, 16, 32, 64, 128, 256, 512, 1,024, 2,048, 4,096, 8,192

and so on. But Malthus reckoned that the available resources, such as food supply, grew more slowly: 'arithmetically', increasing by the same fixed amount after each instant of time. For example,

2, 4, 6, 8, 10, 12, 14, 16, 18, 20, 22, 24, 26, 28, 30, 32

and so on, where each number exceeds the previous one by 2. The modern term is 'linearly'.

Linear growth, even if the number added at each step is very large, will always be beaten in the long run by exponential growth, even if the multiplying factor is only slightly larger than 1. So Malthus deduced that unrestrained growth will always outstrip resources, and concluded that growth always has limits. The argument as presented is simplistic, with its reliance on tidy numerical sequences, but its conclusion is robust. Anything remotely like exponential growth will eventually beat anything that is roughly linear – including growing like some fixed power, squares, cubes, and so on.

Malthus was interested mainly in the human population, but Darwin noticed that the same reasoning must apply to animal populations. Such populations are roughly constant. The number of blackbirds in a given part of the country may fluctuate from year to year, but it doesn't explode. On average, it stays much the same. Yet each pair of blackbirds produces dozens of offspring. What keeps population sizes stable? Clearly, as Malthus said, competition for resources: food, a place to live, mates and, of course, the effects of predators, another kind of competition. The inevitable result would be 'natural selection'. The only creatures that could pass their characters on to the next generation would be those that bred ... and in order to breed, they had to survive to adulthood.

Just as a human breeder might deliberately choose to breed from a faster horse or a thinner dog, so nature would – must – unconsciously 'choose' whichever adult organisms won the competition to survive and breed. And that wasn't a lottery, it wasn't a totally random process. Healthy animals would tend to beat sick ones; not always, but often. Strong animals would tend to beat weak ones. It might not always be obvious to a human scientist which strategy was best (a small animal can hide where a big one can't, for instance), but nature would carry out the experiment automatically and find out what worked.

Here was the missing mechanism. It was well known that different organisms in a given species are not always identical. This process of natural selection would favour certain differences, but suppress others. The result would be gradual changes. How far might such changes lead? Enough small changes, piled on top of one another, can amount to a big one. A very big one. Big enough, Darwin thought, to generate entirely new species – given enough time.

Was there enough time? Darwin's geological background left him in little doubt.

. . . .

How old is the Earth? This may seem a silly question in a book about biology, but even the stoutest supporter of evolution accepts that the process must take a lot of time. Ten thousand years would be woefully inadequate for humans to evolve from an ape-like ancestor – let alone for the ancestor to evolve from fish, and fish to evolve from microbes.

Convince people that the Earth is a mere ten thousand years old, and the battle against what some inhabitants of the US Bible Belt call 'evilution' is won. Evolution *must* be nonsense. So Creationists now routinely dispute the scientifically established figure for the age of the Earth, which is about 4.6 billion years.

Until about 150 years ago, the Bible was one of the main sources of information about the past, so it made sense to try to deduce the date of creation from its contents. James Ussher, Archbishop of Armagh, was an intelligent and capable man with a gift for languages. In 1650 he published the first of two works devoted to Biblical chronology: *Annals of the Old Testament*. A sequel appeared in 1654. In these works he employed Biblical scholarship to work out the date of creation, and the answer he came up with was around nightfall on the day before 23 October 4004 BC.[3]

He was not alone in his endeavours. A decade earlier, John Lightfoot, using similar methods, deduced that the creation occurred near the autumnal equinox in 3929 BC. Isaac Newton, venerated today as one of the principal architects of the Age of Reason, arrived at a date of 4000 BC. Johannes Kepler, famed for his discovery of the laws of planetary motion, proposed 3992 BC. There was a strong consensus that the Earth was about 6,000 years old, so after 1700 the King James Bible included Ussher's chronology among its annotations. It is therefore only to be expected that for several centuries, well-informed Christians *knew* that the Bible stated the Earth's age as 6,000 years.

As the centuries passed, a series of scientific advances provided viable alternatives to Biblical scholarship, and made it possible to date the Earth's rocks objectively, and with increasing accuracy, over increasing periods of time. These dates flatly contradicted Ussher's chronology.

It took a while for the magnitude of deep time to sink in. Initially, ten million years was daring, but soon estimates of the order of a hundred million years became commonplace. There is now a very strong consensus among virtually all scientists that the Earth is close to 4.6 billion years old –*three-quarters of a million times* as old as proposed by the theological chronologists. That makes evolution far more plausible.

Prove that the Earth is young, and evolution is a dead duck. One consequence is Young Earth Creationism, which maintains, contrary to a truly gigantic body of scientific evidence, that the Earth is somewhere between 5,700 and 10,000 years old. Surveys indicate that about 45% of modern American adults accept this figure. They also show that most of those who do are low-paid and poorly educated.[4] Bearing in mind that in the United States it is close to social suicide to admit to agnosticism, let alone atheism, these figures should be neither surprising nor especially discouraging to those of us who accept the science.

. . . .

Armed with deep time and natural selection, Darwin now had the answer to the puzzles in his B Notebook. But as a solid Christian (albeit a Unitarian, a sect often characterised as believing in 'at most one God') he was uncomfortably aware that what he held in his hands was theological dynamite. His wife, Emma, was very religious; he knew she would find his new theories offensive, and he had no wish to upset her. He had no wish to be seen as attacking the Church, either. So he did what most of us would do in such circumstances: he dithered.

He amassed ever more extensive evidence for natural selection. He listed its weaknesses too: Darwin's greatest strength was intellectual honesty. He discussed his ideas with a few trusted colleagues, among them Lyell and Joseph Hooker. In his head, Darwin conceived of a huge multi-volume work, so perfect and so well argued that no sensible person could disagree with it. He might have tinkered and polished and amended and dithered forever, were it not for events taking place half a world away, of which he yet knew nothing.

The agents of those events were a tropical typhoon and a Victorian explorer named Alfred Russel Wallace. Darwin came from a wealthy family. Wallace didn't, and made a living by travelling to distant and exotic parts, collecting butterflies and beetles and other exotic creatures, and selling them. These items were popular with the middle and upper classes in Victorian times, and there were dealers who specialised in them. Wallace went to the Amazon in 1848, and by 1854 he was in Borneo hunting orang-utans. He was beginning to think they might represent human ancestors.

Stuck indoors when a typhoon was dumping rain all over Borneo, Wallace started playing around with a few ideas that had suddenly occurred to him about what he called 'the introduction of species'. He wrote them up as a scientific article and sent it to the *Annals and Magazine of Natural History*. This was a rather unprestigious publication, but Lyell noticed when it published Wallace's paper and told Darwin, since there seemed to be some similarities with the ideas that Darwin was explaining to his friends. Another friend, Edward Blyth, also drew it to Darwin's attention, writing to tell him that he thought the article was 'Good! Upon the whole.' A worried Darwin got hold of a copy, and wrote back in relief to say that it was 'nothing very new ... it seems all creation with him.'

So Darwin relaxed. He had a generous nature, and encouraged Wallace to continue with the work, not realising where that might lead. Wallace followed the well-meant advice, and soon came up with a better idea, essentially identical to Darwin's concept of natural selection. In June 1858 he sent Darwin a twenty-page letter outlining the argument, from which it was immediately obvious that Wallace had come up with a very similar theory, so similar that Darwin declared that his life's work was 'smashed', adding: 'If Wallace had my sketch of 1842, he could not have made a better short abstract!'

Trying to salvage something from the wreckage, Lyell suggested that the two men should publish their discoveries simultaneously, and Wallace agreed. He had no wish to steal Darwin's thunder; he hadn't realised that the great man had been working on anything remotely related. Since he now knew that Darwin had far better evidence, and much more material, Wallace did not wish to steal the limelight.

Spurred on by the worry of being beaten to the punch, Darwin quickly put together a short version of his own work. Hooker and Lyell got the two papers inserted into the schedule of the Linnaean Society. The Society was about to shut up shop for the summer, but its council fitted in an extra meeting at the last minute. The two papers were duly read to an audience of about thirty fellows ... and we've already seen how they were received.

Darwin polished up his essay and changed its title to *On the Origin of Species and Varieties by Means of Natural Selection*. On the advice of his publisher, John Murray, he cut out the 'and Varieties'.

The first print run of 1,250 copies went on sale in November 1859 and sold out immediately.

The Wallace–Darwin theory of natural selection is simple – deceptively so, as will become apparent. The main points are:

1. living creatures differ from one another, even within a given species;

2. many of these differences can be passed on to offspring;

3. the environment cannot sustain all offspring produced, so there is competition for survival;

4. the survivors tend to be 'better' at surviving than the previous generation.

The deduction is that species can gradually change; and given enough time, small changes can combine into large ones.

This, said Darwin, is how new species arise from existing ones.

. . . .

Natural selection is deceptively simple: most attempts to explain the ideas in non-technical terms, including my own attempt here, are forced to simplify a very complex process. Often they oversimplify it. A classic instance is Herbert Spencer's characterisation of natural selection as 'survival of the fittest'. Another is 'nature red in tooth and claw', from Tennyson's poem 'In Memoriam A.H.H.',[5] which actually referred to humanity, but was quickly seized upon by both proponents and opponents of evolution. A third is the idea, often stated by biologists as a flat fact, that evolution is random.

Spencer coined his memorable phrase in 1864, in his *Principles of Biology*. Biologically, 'fitness' is about how well an organism is equipped to 'fit into' its environment, but many people assumed it was about being healthy and in good physical shape. In addition, Spencer's vision was closer to Lamarckism, so the book muddied already turbid waters. Critics often cite the phrase 'survival of the fittest' as proof that evolution is a 'tautology': survival is used to demonstrate fitness, and then fitness is proposed as the reason for survival.

There are two mistakes here. First, a tautology is a logical statement that is unconditionally *true*; the correct objection should be to circular reasoning. For example, it is fallacious to argue that the existence of life implies that there must be some supernatural 'living essence', and then use that essence to explain life. The more serious mistake is the assumption that Spencer's dumbed-down phrase accurately describes the technical concept of natural selection. This misleadingly suggests that biologists think that creatures have an innate fitness, and the fitter creature wins. But evolutionary biologists do not use survival as a demonstration of fitness and then use fitness to justify survival. What matters is *selection*: since creatures compete for limited resources, some survive and some don't. The process is a kind of filter, and what matters is that some organisms pass through it successfully, while others don't. And filters do not just allow things to pass through, or block them, at random: there is a degree of systematic bias. If we don't understand in great detail how the filter works, we may not be able to predict what will pass through and what won't, but we can still predict that the distinction will be systematic.[6] Organisms don't compare their fitnesses to determine a winner. They just compete, and find out which one wins.

The word 'compete' also bears examination. Think of a wood, populated by foxes, rabbits and owls. Which competitions are the most significant for evolution? 'Nature red in tooth and claw' leads us to home in on red-toothed foxes, or predatory owls hunting down innocent rabbits: the obvious competition is between predator and prey. But which organisms constitute a rabbit's most serious competition?

Other rabbits.

Rabbits are in competition with one another for the same resources, and in the struggle to survive the same dangers. They are competing for the same food, and trying to escape from the same foxes and owls.[7] The rabbits that win these ongoing competitions are the ones that will survive to breed, and the abilities that enable them to do so (if they are of a kind that can be inherited) may pass to their descendants and similarly improve their survival prospects. The same goes for foxes, whose main competitors are other foxes, and for owls, whose main competitors are other owls.

Competition between foxes and rabbits happens as well, and it also has an evolutionary effect, but on a slower timescale. It leads to an 'arms race' in which, say, rabbits acquire the ability to run faster, but foxes evolve to do the same, and the whole cycle goes round and round, driving both creatures to ever greater efforts.

However, there is a third competition going on here: it is between foxes and owls. Foxes don't usually eat owls, and owls don't eat foxes – well, maybe baby ones. But they both eat rabbits, and if the rabbit supply gets too low, owls and foxes can't avoid coming up against each other. This competition is indirect; the participants may not even be aware that their competitor exists. All they know is that there aren't enough rabbits around and they're going hungry.

Natural selection is not merely a matter of pitting one organism against another. It happens in the context of the surrounding environment – the entire local ecosystem. The rabbit population depends on the plants they eat, the availability of suitable soil to dig burrows, the amount of low cover to hide in. These things depend on other, less obvious features of the ecology, such as insects to fertilise the plants, and fungi and bacteria to condition the soil. So what evolves is really the entire ecosystem. Yes, the evolution is driven by what happens to individual organisms, but the ecosystem determines the context in which they compete. Mathematical models discussed in Chapter 14 underline this point.

Is evolution random? It's clearly not predetermined – you can no more forecast which rabbit will make it through the day than you can predict whether France will beat Mexico in the World Cup. But there are clear trends, some events are more likely than others, and so on. Down at the molecular level, genetic changes can perhaps be characterised as random. Mutations, genetic changes, can be caused by chemicals and cosmic rays, which are effectively random.[8] But that doesn't imply that evolution itself is equally likely to change in one direction as it is in another, and the reason, yet again, is natural selection. Depending on the appropriate context, selection introduces a degree of preference. Some mutations improve the creature's survival chances, some decrease them, and most are neutral.

Let me suggest an analogy. The motion of individual molecules in water is random, but that doesn't mean that water is just as likely to flow uphill as down. The selective effect of gravity leads to a strong preference for down. But what the water does, in bulk, is not merely determined by moving downwards. Where it ends up depends on the landscape in which it flows. Natural selection is rather like the force of gravity (though not as strongly selective and not as easy to characterise independently of its effects). Mutations are like the random excursions of water molecules. And the environment is like the landscape.

. . . .

In 1973 Theodosius Dobzhansky, a prominent evolutionary biologist, wrote an essay with the title 'Nothing in biology makes sense except in the light of evolution'.[9] Everything that has been discovered in biology since then – which is an awful lot – has reinforced that statement. The word 'theory' has two very different meanings, and there is an overwhelming scientific consensus that in 'theory of evolution' this word has made the transition from the sense of 'tentative hypothesis' to the sense of 'coherent explanation confirmed by a substantial body of evidence from diverse sources which has survived innumerable attempts to disprove it'. As Richard Dawkins has remarked, the everyday term for this sense of 'theory' is 'fact', and it is mainly a wish to avoid appearing dogmatic that prevents scientists from employing the same term.

It would hardly be necessary to point this out, were it not for vocal opposition to evolution by a few fundamentalist religious groups. If the world was indeed created by God ten thousand years ago, then the Deity has gone to enormous lengths to fabricate a massive, interlocking network of natural features, specifically designed to mislead any intelligent observer into the mistaken belief that life on Earth has diversified, over billions of years, from simple beginnings. This view of God as liar seems theologically improper, and that is the conclusion that Victorian clergy came to once they had absorbed the scientific discoveries of their age. So did Dobzhansky, a Russian Orthodox Christian.

The evidence for evolution comes from many different sources. The variety of these sources, and their independence from one another, greatly strengthens the scientific case in favour of evolution, because each new source provides a large number of potential ways for the theory to be disproved. So far the basic principle has survived unscathed, but the details of the evolutionary process have been clarified, and sometimes changed, as new evidence comes in. The evidence available today is far more extensive than what was available in Darwin's time; it is also more quantitative and more precise. The main sources are:

- the pliability of the form and behaviour of organisms, evident in human-induced breeding programmes for dogs, pigeons, horses and other domesticated animals;
- similarities between existing creatures, suggesting a common origin;
- the occurrence of the same biochemical components and systems in many different organisms;
- the fossil record, which reveals coherent sequences of changes over time;
- the geological record, which confirms the dating of fossil species;
- genetic features of organisms, especially DNA sequences, which confirm both lines of descent and the timing of changes;
- relations between the distribution of species and current or historical geographical features;
- observed natural selection in the laboratory and the real world;
- mathematical studies of the effect of selection principles on changes in complex systems.

. . . .

Evolution's critics often claim that because we can't observe the past, the theory is not scientifically testable. But science is about inference as well as direct observation. When Haldane was asked what evidence could possibly disprove evolution, his immediate response was: 'Fossil rabbits in the Cretaceous'. Fossils are relics of living creatures from the past, transformed into and preserved in the rocks of our planet. Because rocks are often datable, many fossils can be reliably assigned to specific periods of history.

Fossils provide a rather sparse record of the life forms of the past, because it is very rare for any individual organism to become a fossil. However, there have been an awful lot of organisms over the past few hundred million years, and more than 250,000 different fossil species have been discovered. The number of individual known fossils is large (more than three million just from the La Brea tar pits at Los Angeles, for example) and is increasing rapidly with the discovery of new sites, all over the world, and with improved techniques for locating fossils and analysing them.

Despite the relative rarity of fossils, the record is sometimes very extensive, with few significant gaps, and it then offers clear evidence of systematic long-term evolutionary changes. The classic example is the evolution of the horse, between 54 million years ago and a million years ago. The sequence begins with a horse-like mammal a mere 0.4 metres long. This genus was originally given the poetic name *Eohippus* ('dawn horse'), but has since been renamed *Hyracotherium* because of the rules of taxonomy, which in this case managed to deliver a silly result.[10] The sequence continues with *Mesohippus*, 35 million years old and 0.6 metres long; then *Merychippus*, 15 million years old and 1 metre long; then *Pliohippus*, 8 million years old and 1.3 metres long; and finally (so far) *Equus*, essentially the same as the modern horse, 1 million years old and 1.6 metres long.

Taxonomists can track, in great detail, the sequence of changes that occurred in this lineage of ancient horse ancestors, for example in the animal's teeth and hooves. They can also track the timing of these changes, because rocks can be dated. So now evidence from geology can be thrown into the mix. In principle it would take only one fossil species in the wrong stratum of rock to cast doubt on the evolutionary story – and that is Haldane's point. In practice it would take several independent instances, because there might exist sensible explanations for a few isolated exceptions. The plain fact is that the succession of rocks, their ages as determined by a variety of different methods, and the evolutionary sequences of fossils all agree to a remarkable extent.

A standard objection to the use of fossils to support evolution is the absence of transitional forms, popularly known as 'missing links'. Neither term is terribly satisfactory: in evolutionary biology there is a sense in which *all* species are in transition (from their ancestral species to their descendants), and transitional forms are not modern 'links' but ancient common ancestors. However, the important question is whether these forms are absent because they never existed, or because they did exist but we haven't yet found any fossils. Evolution predicts the latter – yet another way in which evolution has predictive power – while its detractors assert the former.

As more and more fossils have been discovered, more and more transitional forms have been found. A major example is the transition from fish to land animals – tetrapods, creatures having four limbs instead of fins. The only fossils available to Victorian palaeontologists were either fish or amphibians, leaving a gap of at least 50 million years with no transitional fossils in between. But from 1881 onwards, new fossil discoveries have inserted a whole series of intermediates between fish and amphibians: *Osteolepis*, *Eusthenopteron*, *Panderichthys*, *Tiktaalik*, *Elginerpeton*, *Obruchevichthys*, *Ventastega*, *Acanthostega*, *Ichthyostega*, *Hynerpeton*, *Tulerpeton*, *Pederpes*, *Eryops*. Dozens of other comparable gaps have been filled in the past twenty years, and every year now sees the discovery of more transitional forms, ever more finely spaced.

One evolutionary critic has remarked that whenever a gap in the fossil record is filled, it creates two more gaps on either side of it. This rather desperate excuse is true in a trivial sense, but it constitutes a serious misunderstanding of the nature of scientific inference. Each discovery of another transitional form represents a successful prediction for evolution and a failure for its detractors. Moreover, the two gaps that are created are significantly smaller than the one that was there before. A new transitional form is one more nail in the coffin of special creation. Enough nails will fix the lid on firmly; it is not necessary to have an unbroken continuum of them all round the edge.

. . . .

It wasn't his wish to create controversy, but Darwin had opened Pandora's box. As the implications of the *Origin* struck home, and even more so those of its successor, *The Descent of Man*, which argued that humans and apes are descended from a common ancestor, hackles were raised. Sensitivities were trampled. Social convention was outraged.

If your default worldview places humankind at the pinnacle of creation, with the rest of the universe as a resource for us to exploit, then the suggestion that people and animals have a lot in common is hard to accept, and the idea that humans and today's animals evolved from common ancestors is anathema. It inevitably led to snide remarks like, 'Exactly which ape was your grandfather, Mr Darwin?', and unflattering cartoons, many of which would now be classed as racist.

From a more dispassionate viewpoint, however, the close relationship between humans and animals is obvious. We eat like animals, reproduce like animals, excrete like animals. Our anatomical features correspond closely to those of many other members of the animal kingdom. Our skeletons match those of most mammals virtually bone for bone; only the shapes and sizes are a bit different. Our brains have a lot in common with those of mammals, amphibians and reptiles. Our hands are very similar to those of the great apes, and not so different from those of monkeys and lemurs.

When a difference of opinion on such matters was just that, the outcome was pre-empted by social convention. In the predominantly Christian tradition of the Western world, it was taken for granted that people are completely different from animals. The differences, such as our ability to talk, write, compose music or paint portraits, were emphasised; the similarities (especially those related to embarrassing bodily functions) were ignored, minimised or, as a last resort, denied. But as scientific evidence accumulated, this stance became difficult to maintain. In Victorian England, and most of Europe, religious people slowly came round to the idea that the creation story in Genesis is a metaphor, and they accepted the discoveries of science as insights into God's creation. Atheists had no problem anyway. But those who believed in the literal truth of the Bible painted themselves into an intellectual corner, by tying their entire belief system to an unconvincing denial of a huge and ever-growing body of scientific evidence.

6   In a Monastery Garden

Today's academic scientists live or die by their citation ratings –
how many other scientists have referred to their papers in
published research. Bureaucrats love citations – like paper clips, you
can count them. However, there are dangers. In mathematics, some
of the best papers are so well known that no one bothers to
mention them explicitly. But the biggest problem is the time it can
take for the importance of a discovery to become apparent. A case
in point is a paper published in the nineteenth century that created
the entire subject of genetics. The discoveries and ideas that it put
forward have proved absolutely fundamental to our understanding
of living creatures, yet in the 35 years after it appeared in print, it
was cited no more than three or four times.

The paper, written in German, was published in 1865 in an
obscure journal, the *Verhandlungen des Naturforschenden Vereines in
Brünn* ('Proceedings of the Natural History Society of Brünn'). The
author, born in Germany, was christened Johann. As a child he
kept bees and worked as a gardener. In 1840 he became a student at
the Philosophical Institute of Olomouc, a city in Moravia, part of
today's Czech Republic. After a single term at the Institute he fell ill
and took a year out. After finishing his studies, Johann decided to
become an Augustinian priest. He changed his first name to the one
he would use in monastic life: Gregor. His surname was Mendel.

In 1851 the Order sent Mendel to the University of Vienna, and
on his return to the abbey he became a teacher. There, in 1856, he
began a series of 29,000 scientific experiments, breeding peas. It
took him seven years. After peas, he moved on to bees, but with
less success. He bred a strain of bees that had to be destroyed

because they were so nasty, and he failed to get clear-cut results because it was difficult to control the queen bees' choice of mates. In 1868 he was promoted to abbot, and his scientific productivity ceased. But what he achieved would eventually trigger biology's fourth revolution: genetics.

It was a struggle. Most biologists of the day rejected Mendel's theories, mainly because they conflicted with the prevalent belief that characters passed from parents to offspring by 'blending'. The main idea here – if it can be dignified by that word – is that a child will have a height that is somewhere in between the heights of its parents, as if the two heights were poured into a mixing-bowl, stirred together and poured into the child. Height can be replaced by any other character: weight, strength, size of biceps, mathematical talent, whatever.

Evidence in support of the blending theory of inheritance was thin on the ground, while contrary evidence – most of it blindingly obvious – was widespread. Nevertheless, virtually everyone believed in blending inheritance. I suspect that one motivation was the then-popular metaphor of 'blood' for inherited characters. Animal breeders would refer to 'bloodlines' to describe the family trees of dogs or horses. Even today we speak of someone having 'royal blood', or being a 'blood relative'. This metaphor can be traced back to ancient Greece, and became known as pangenesis (*pan*=whole, *genesis*=birth, origin). Even Darwin fell into the trap: when he wrote the *Origin*, it was pangenesis that he had in mind as the mechanism of heredity.

However, blending inheritance makes no sense, as became apparent once blending was confronted by science. Between 1869 and 1871, Darwin's cousin Francis Galton, one of the pioneers of statistics, performed a long series of experiments to test the theory of pangenesis. His approach was disarmingly direct: he transfused blood from various types of rabbit into other types, then he bred them and observed the characters of the resulting offspring. He found no indication of any substance in a rabbit's blood that determined its offspring's characters, and pangenesis was rapidly abandoned by most competent biologists. But before Galton, pangenesis was simply *there* – a cloud of unstated and unquestioned assumptions floating around in the heads of biologists, breeders and the general public. If you were really clever, you could find cunning ways to prop it up, just as an experienced flat-earther can always win a debate, point by point, by invoking unorthodox theories of optical refraction, weird geometry – or, when desperate, conspiracies.

Against this background of unquestioning acceptance of pangenesis, Mendel's results stood out like a sore thumb. But instead of trying to understand them, or repeating and extending his experiments, it was so much simpler to ignore them – assuming you had even read the paper. Darwin hadn't. If he had known of Mendel's work when he wrote the *Origin*, he would have made some big changes.

* * * * * * * * * * * *

At first sight, Mendel's experiments do not appear terribly revolutionary. All he did was breed pea plants and compare the characters of the new generation with those of the previous one. But what he found was potentially explosive, and eventually, after his death, it detonated with a bang that can still be heard – at least, by anyone who doesn't stuff their head with nonsense in the hope that it will plug their ears.

Mendel's paper languished, unread and unappreciated, until about 1890, thirty years after the publication of the *Origin*. It was rediscovered by two botanists, Hugo de Vries and Carl Correns.

Mendel's discoveries hinge on some simple numerical relationships that he observed when breeding pea plants. The basic idea was straightforward: focus on various specific characters, cross-fertilise a plant that has a particular version of that character with another plant, one that has either the same version or a different one, and see what the corresponding character is in the next generation. Cross-fertilisation, or cross-breeding, means that pollen from one plant (I'll call this the 'father') is used to fertilise the other one (the 'mother'). Plants are ideal for this kind of experiment, because the scientist can paint pollen from the father directly onto the reproductive organs of the mother, which makes it easy to control the line of descent. Not so easy in angry bees!

One of the first characters that Mendel studied was the colour of the flower: white or purple. The first thing that struck him was that these were the only colours that appeared. There were no signs of blending, no pale purple or purplish-white flowers. No matter how many times he cross-bred the pea plants, their flowers remained resolutely either white or purple. The theory of blending inheritance didn't fit the evidence, so Mendel set out to discover what really happened.

It might seem obvious that if you cross two 'white' plants – that is, ones with white flowers – you should always get white plants, and ditto for purple. But that assumption smacks of blending, and it's wrong. Mendel found that white plus white always gave white, but purple plus purple could give either colour. So it wasn't a simple case of two distinct 'races' of pea plants, whose offspring were the same colour as the parents. It was more complicated. In fact, there seemed to be *three* different outcomes when crossing purple with purple:

- all the offspring are purple;
- three-quarters of the offspring are purple and the other quarter are white; or
- half the offspring are purple and half are white.

In contrast, a white–purple cross could behave like the first two of these possibilities, but the third, and intuitively the most natural, didn't happen. The proportions of different outcomes – a half, a quarter, three-quarters – weren't exact; they varied from one experiment to the next. But the observed data fitted these proportions well.[1]

What was going on? An important step towards the answer is to select the plants you cross-breed, which simplifies the possible results. Say that a particular character 'breeds true' if it reappears in all offspring. Breeding true depends on *both* parents, but by storing some of their seeds and using the others to grow a new generation, and then cross-breeding those, you can sort out which seeds came from plants that bred true, and use those plants' remaining seeds in another experiment.

It now turns out that if you cross a pure-bred white plant with a pure-bred purple one, the result is *always* purple. However, if you pick two of the plants from that new generation, and cross-breed those, then you *always* get roughly three-quarters purple and one-quarter white in the succeeding generation. This is bizarre – it's almost as if the plants have some sort of 'memory' of past generations. And in a sense, they do.

You can imagine poor Mendel, puzzling over his observations, trying to find a sensible explanation. Eventually he realised that everything made sense if the character 'colour' was determined not by one genetic factor in any given plant, but by two. One factor would be inherited from the father, the other from the mother. What these factors were, physically, was a mystery. But the numbers, the mathematical patterns, strongly suggested that they must exist.

Suppose that colour is determined by unspecified factors that can be either W or P – white or purple – and that each plant has two of them. The possible pairs are WW, WP and PP. We consider PW to be the same as WP: what counts is the combination of factors, not the order in which they are written down.[2]

When two plants are cross-bred, the offspring inherits one factor from each parent. If both factors are identical – WW or PP – then it makes no difference which of the two is inherited. These are the 'true-breeding' plants. But suppose that WP breeds with, say, PP. Then the offspring can inherit either W or P from the first parent, but must get a P from the second. So there are two outcomes: WP or PP.

The mathematics involved here is combinatorics: how different mathematical objects can combine – here, the symbols W and P. But in this case you don't need to know any combinatorics to figure out the answer using 'bare hands':

- If we cross-breed WW with WW, then the only possibility is WW.
- If we cross-breed PP with PP, then the only possibility is PP.
- If we cross-breed WW with PP, then the only possibility is WP.
- If we cross-breed WW with WP, then there are two possibilities: WW and WP.
- If we cross-breed PP with WP, then there are two possibilities: PW (=WP) and PP.
- If we cross-breed WP with WP, then there are four possibilities: WW, WP, PW and PP. But PW=WP, so the four possibilities reduce to three.

What about the proportions that Mendel observed? Those clinch the argument. To see why, it helps to draw a diagram, known as a Punnett square after the British geneticist Reginald Punnett, who invented it around 1900. I'll look at WP and WP; this is one of the most complicated cases, but more typical and therefore easier to understand.

The top row in Figure 13 shows the two factors (W and P) present in the mother; the left column shows the two factors (again W and P) present in the father. The four squares show the resulting combinations (WW, WP, PW, PP) when particular factors are present in the offspring. The usual convention is to put the factor derived from the father first. We've seen that the order doesn't affect the character of the resulting plant, but it helps to keep the mathematics straight.

Fig 13   Punnett square showing how WP cross-breeds with WP.

I've coloured some of the big squares white and others grey: these represent the colours of the flowers of the corresponding plants, with grey standing for purple. I've also broken with tradition by attaching rectangular tags in the top corner: these represent the colours of the parents. The shading tells us that WW gives white, whereas WP, PW and PP give purple. The idea – very simple, like all good ideas, and one of Mendel's great insights – is that W and P 'vote' on the colour, but if W tries to contradict P, then P wins. In the genetic jargon, W is recessive and P is dominant.

It is this voting rule that makes mixed cases like WP select one of the two colours found in the parents, instead of somehow blending them, or doing something else. In principle, the 'purple wins' voting rule is just one possible way to assign a colour to a plant with mixed factors; many others can be conceived. This method is very neat and simple, and it works for the colours of pea plants. However, biology being what it is, the more geneticists investigated such rules, the more alternatives they discovered. Ironically, some amount to blending.

In Figure 13 three of the squares are grey (purple flowers) and only one is white (white flowers). This 3:1 proportion of purple to white is exactly what Mendel found in some of his experiments. It suggests that the numerical regularities Mendel observed in the proportions of plants with various characters must have a statistical explanation. The numbers are evidence about the *probabilities* of various outcomes.

Now another area of mathematics has joined the party, alongside combinatorics: probability theory. This is one of the major branches of the subject, the mathematics of uncertainty. It originated in questions about gambling; the first textbook was Jacob Bernoulli's *Ars Conjectandi* in 1713. I like to translate this as 'The Art of Guesswork', but a more faithful translation is 'The Art of Conjecture'. Bernoulli defined the probability of some event to be the proportion of times that it happens, in the long run, over large numbers of trials. This fits with intuition. For example, if we roll a fair die, then each face – 1, 2, 3, 4, 5, 6 – 'ought to' come up roughly the same number of times. If 6 kept turning up more often than 2, the die wouldn't be fair.

This is fine as a working definition, but it entails an assumption: that what happens in the long run is representative. However, it is certainly *possible* to throw a hundred 6's in succession with a fair die. Bernoulli proved a mathematical theorem, the law of large numbers, which shows that exceptions of this kind are extremely unlikely. Later, mathematicians put the whole subject on a sound logical basis by stating an explicit list of *axioms*: properties that any notion of probability must satisfy. The law of large numbers then becomes a theorem, a logical consequence of the axioms, and it lets us calculate probabilities combinatorially – by counting. So we can calculate the probability of a purple flower by counting how many combinations of factors give purple, and dividing by the total number of combinations: here 3 divided by 4.

Mendel's scheme for heredity combines characters from both parents while avoiding blending. It treats both father and mother in the same way. The father has two factors, but contributes only one to the offspring; ditto for the mother. In each case we have to choose one factor from two. Suppose this is done at random, just like tossing a coin: heads, one factor, tails the other. This implies that each factor from the father is equally likely to be chosen, and similarly for the mother. So each separate combination in the Punnett square is equally likely, having probability 1/4. Since there are three grey regions out of a total of four, we expect 3/4 of the plants to be purple. Since there is only one white region, the remaining 1/4 should be white. So the combinatorics of the two symbols W and P, subject to the voting rule 'P wins if present', represents the observed frequencies of the two colours – provided we choose one factor from each parent at random with equal probabilities.

Similar calculations explain the proportions that Mendel observed in other cases. The random aspect of the process explains why Mendel's observed proportions were not *exact* fractions like 3/4 and 1/4. In random processes there is always a degree of 'scatter', when things don't behave exactly like the average case, which is what the probabilities reflect. For example, if you toss a coin four times in a row, then the 'average' or 'expected' result is two heads and two tails. However, the actual result may be anything from four heads to four tails, and the average case happens less than half the time.

Mendel didn't stop when he had his great insight. He devised methods to test this hypothesis. The trick was to remove the annoying plants that produced different colours, by breeding several generations and discarding any plants whose offspring were not all the same colour. Having identified particular plants that bred true, Mendel could go back to his store of seeds, and use the seeds from those plants to grow new ones which he could then cross-breed in various ways. After a few generations had passed, clear patterns set in, and they supported his theories.

· · · · · · · · · · · · ·

To Mendel, genes were mysterious 'factors', and he did not know where they were located in the organism, or what they were. The answer emerged from studies of cell division. A cell is not a simple bag of chemicals, but a highly complex, organised structure – organised enough, and complex enough, to reproduce. It's an amazing trick to copy a cell, but that pales into insignificance compared with the copying of an entire organism. This process, fundamental to complex life, has piggybacked itself on a special kind of copying process for cells.

Prokaryotes reproduce by splitting into two copies: this process is called binary fission. Eukaryotes also split into two copies, but because such cells are more complex, their division is also more complex. Additionally, eukaryotes are usually capable of sexual reproduction, in which the offspring has genetic contributions from two (or for a few organisms like yeast, possibly more) parents; Mendel's pea plants are an example. For sexual species, creation of the relevant germ cells (sperm and eggs) involves this second kind of cell-division, called meiosis.

For a long time after Mendel had inferred the presence of genetic 'factors' in plants, no one knew the physical (that is, we now realise, molecular) basis of heredity. When artificial dyes became available, it was discovered that thin sections of cells could be stained to reveal hidden structures under the microscope. Among them were puzzling features known as chromosomes – coloured bodies. Prokaryotes had a single chromosome, forming a loop attached to the cell wall. Eukaryotes kept their chromosomes inside the cell nucleus, and each organism had a particular number of chromosomes – 46 in humans, for instance. The chromosomes were shaped roughly like an X, and came in many different shapes and sizes.

Chromosomes were somehow involved in cell division, because an early step in the division of both prokaryotes and eukaryotes involved making copies of them. With this as a clue, biologists began to suspect that chromosomes were the cell's genetic material. Theodor Boveri and Walter Sutton independently came up with this idea in 1902, and performed a series of experiments to test it. Boveri worked with sea urchins, and showed that unless all chromosomes were present, the organism failed to develop correctly. Sutton focused on grasshoppers, and made the crucial discovery that chromosomes came in pairs, one member of each pair derived from the father, the other from the mother. These pairs surely must be Mendel's factors.

This proposal remained controversial for about ten years, but in 1913 Eleanor Carruthers showed that chromosomes combined together independently, which was consistent with the numerical ratios that Mendel had observed. For example, the 46 chromosomes in a human come in 23 pairs, but germ cells contain only one member from each pair (see later). This comes from either the father or the mother, and the choice is made randomly and independently for each pair. The clincher came two years later, when Thomas Hunt Morgan carried out definitive experiments on the fruit fly *Drosophila melanogaster*. He showed that genes associated with regions of the chromosome that are very close together tend to be associated in descendants: either they have both or they have neither. This biasing effect slowly weakens as the regions get further apart.

In the binary fission of a prokaryote, the first step is to make a copy of the single loop-shaped chromosome. After that, the cell grows in size. The two copies of the chromosome attach themselves to the cell membrane. Then the cell grows longer, separating the chromosomes. Finally, the cell membrane grows inwards, eventually splitting the cell so that the chromosomes end up in distinct halves. The end result is two copies of the original cell, more or less identical to it, and in particular having the same genetics. (This is not quite true, because copying errors can occur, but I'll leave that for later.)

. . . .

The reproduction of a eukaryote cell is more complicated, and is known as cell division. It can happen in two different ways: mitosis, in which the daughter cells are also able to reproduce, and meiosis, in which they turn into gametes, the basic units of sexual reproduction. In humans, these are sperm cells in the male and ova (eggs) in the female.

Mitosis begins in the nucleus of the cell. The first step, again, is to make a spare copy of the cell's genetic material. In eukaryotes this is packaged into several chromosomes, so each chromosome must be copied. This is generally done for all the chromosomes at the same time, rather than taking them in turn. Then the pairs of chromosomes are pulled apart into two sets, each containing one chromosome from each identical pair, while the nucleus divides into two parts, each containing one set of chromosomes. While this is going on, the cell's component organelles, such as mitochondria, are also duplicated, by processes that closely resemble binary fission in prokaryotes. Finally, the cell membrane grows inwards and splits, in a way that ensures that each daughter cell contains its fair share of all of these components – in particular, one nucleus.

This sequence is typical but not unique: the details of mitosis are different in different organisms. Mitosis is carefully choreographed; biologists distinguish five successive stages (see Figure 14). The mother cell's duplicated contents must be sorted into two separate sets. The dividing cell does this using microtubules, long molecules that normally form the cell's

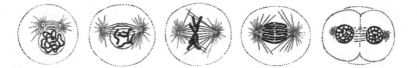

Fig 14   Stages in mitosis. *Left to right*: Prophase: centrosome splits.
Prometaphase: microtubules enter the nucleus. Metaphase: chromosomes
align at right angles to microtubules. Anaphase: microtubules start to
shrink, pulling pairs of chromosomes apart. Telophase: sets of
chromosomes are collected in two nuclei, cell membrane starts to cleave.

'skeleton' and act like ropes that can winch the various organelles
into their correct positions.

Each organelle behaves rather like a prokaryote; in particular, it
reproduces by binary fission. This provides a clue to the origin of
eukaryote cells: they are, to some extent, colonies of once separate
prokaryotes, which have evolved to cooperate inside a larger unit,
the eukaryote cell. This idea is called the endosymbiotic theory. It
was first proposed in 1905 by the Russian Konstantin
Mereschkowski, who pointed out that the chloroplasts in plants,
which contain their chlorophyll, divide in a manner that is
strikingly similar to the division of cyanobacteria, which are
prokaryotes. In the 1920s, Ivan Wallin made a similar proposal for
mitochondria. These suggestions found little favour until the 1950s,
when it was discovered that these and other organelles contained
their own DNA, separate from the main genome of the cell. In 1967
Lynn Margulis provided further evidence for the idea that eukaryote
cells arose as a kind of symbiosis among many different
prokaryotes, incorporated into the evolving cell in a series of steps.

• • • • • • • • • • • •

Prokaryote reproduction is refreshingly direct: an organism divides
into two organisms. In eukaryotes, the reproduction even of *cells* is
less direct, and the reproduction of organisms is very indirect.
Eukaryotes make two copies of the genetic information in certain
cells of the organism, and then build a new organism from scratch
using that information. Reproducing a prokaryote is like breaking a
piece of chalk in half to get two pieces of chalk. Reproducing a
eukaryote is like making a blueprint of a car, photocopying the
blueprint and using that copy to manufacture a new car – with the
extra twist that the blueprint was stored in the glove compartment

of the original car, and its copy is placed in the glove compartment of the new car.

The process that initiates the copying of genetic information is meiosis. This follows roughly similar lines to mitosis, but it has eleven stages instead of five. The most important difference is that the chromosomes are not duplicated, but split apart. In organisms that reproduce sexually, the chromosomes normally come in pairs – one inherited from the father, one from the mother. In meiosis these pairs randomly swap their genetic material, a process known as recombination and the main source of genetic variation within a population. The modified pairs are separated. The end result is a set of four cells, each containing half of the normal complement of chromosomes.

Unlike mitosis, meiosis is not a cycle – at least, not in a single organism. It creates germ cells, and having done so, it stops. The germ cells do very little until two adult organisms, of opposite sexes, do what comes naturally and fertilise an ovum with a sperm. At this point, the two half-sets of chromosomes reconstitute a complete set. The fertilised egg starts to develop, and grows to form the juvenile stage of the same type of organism. In short, two adults have a baby.

If you think of an organism as a cake, then mitosis cuts the cake into two pieces. Meiosis copies the recipe for the cake and tucks it away in a drawer, to be used when required to bake a new cake. But these cakes can grow, and the recipe is tucked away inside the cake.

Because meiosis involves recombination, the child's genome is a mixture of the genomes of its parents – part random, part systematic. In humans, the child is (normally) endowed with the correct 23 pairs of chromosomes. Each pair consists of one chromosome from the father and a corresponding one from the mother. One member is donated by the sperm, the other by the egg. In 22 of these pairs the two chromosomes concerned have the same overall structure, the same sequence of 'genes', but they may differ in the choices made for any particular gene. For instance, human hair comes in a variety of colours: brown, black, auburn, and so on. The colours are caused by pigment proteins called eumelanin and pheomelanin. Eumelanin can occur in two forms: brown and black. Pheomelanin is pink or red. Proteins are made by genes, and different choices of the appropriate genes lead to different colours of hair.

Considering how obvious hair colour is, and its long-recognised relation to heredity ('she's got her mother's hair ...') we don't yet know precisely which genes determine hair colour or how they do it. A popular theory is that there are two genes: one in which brown is dominant and blond is recessive (like purple/white for peas), and another in which suppressing red colour is dominant and red is recessive. But it doesn't explain the full variety of human hair colours.

The 23rd chromosome pair is different: it contains the sex chromosomes, which determine the sex of the child. In mammals (and more) the female sex chromosome X is much larger than the male Y. Females have the pair XX, males the pair XY. The presence of two X's ensures that the set-up is stable under reproduction: the same possible pairs XY and XX are repeated in the offspring, because the child must get an X from its mother. Errors can occur; in particular, a child may have three sex chromosomes rather than the normal two.

One member of each pair of chromosomes comes from the father, the other from the mother, with possible genetic differences. This is the molecular explanation for Mendel's observation that the only sensible way to explain the results of his experiments was to assume that any given character resulted from two 'factors', one from each parent. This process offers one way to combine genetic 'information' in new ways, while retaining its overall organisation: it allows reproduction without exact replication, providing a source of genetic diversity. This in turn opens the door to evolution – in fact, it makes some kind of selective filtering of organisms pretty much inevitable, since it provides a source of heritable variation.

The most intriguing feature of this process, however, is that there is a second, more drastic, source of genetic variation: recombination. A sperm cell from the father does not contain one or other of his chromosome pairs. If it did, then it would be either a copy of the corresponding chromosome from *his* father, or the one from his mother. Instead, they contain a jumbled-up crossover of both pairs: part of his father's chromosome, with the gaps filled by the complementary pieces from his mother's.

Without recombination, separating chromosome pairs into halves and then putting two halves together – one from the father, one from the mother – would be a way to change how chromosomes are paired, but it wouldn't alter the genetic information inside individual chromosomes. This would be a rather feeble way to mix up the genetics. Recombination means that genes get modified within chromosomes, a far more drastic way to alter the genetic make-up. A curious consequence of this two-step mixing process is that the most significant differences between the child's genes and those of its parents arise by jumbling up what the parents inherited from the *grandparents*.

7   The Molecule of Life

. . . .

DNA is now a cultural icon. Scarcely a day goes by without some
company or person claiming on television or in the newspapers
that some activity or product is 'in their DNA'. It is hailed as 'the
molecule of life' and 'the information needed to make an
organism'. It is linked to new cures for diseases, the 'Out of Africa'
theory of the spread of early humanity across the globe, and
whether, tens of thousands of years ago, some ancestors of modern
humans had sex with Neanderthals (they did).

At times, DNA seems to be accorded almost mystical
significance. We are often told that someone is having a child 'in
order to pass on his (or her) DNA', or perhaps 'to pass on their
genes'. Maybe today some people genuinely have children for those
reasons, but for hundreds of thousands of years people have been
having children because they wanted to, for personal reasons, or
because they couldn't avoid it. Their genes got passed on anyway,
along with their DNA, as a vital part of the process ... but that
wasn't the *reason*. It can't have been. They didn't know they had
genes.

Passing on genes or DNA can be viewed as an evolutionary
reason for having children, one that helped instil in us a strong
drive to reproduce despite the dangers of childbirth, but we should
not confuse human volition with the mechanistic workings of
biological development and natural selection. That confusion is a
symptom of DNA as icon, something with almost magical
significance. But after decades of telling the public that working out
the human DNA sequence would lead to cures for innumerable
diseases, or allow scientists to engineer new creatures, biologists

should hardly be surprised when members of the public take these claims seriously.

There is no doubt that DNA is important. The discovery of its remarkable molecular structure was probably the biggest scientific breakthrough of recent times. But DNA is only one part of a far more complex story. And however magical it may seem, it doesn't work by magic.

. . . .

Humanity's tortuous path to the chemical nature of the gene, and the beautiful geometry of the molecular carrier of heredity, took more than a century.

In 1869 a Swiss doctor, Friedrich Miescher, was engaged in a very unglamorous piece of medical research: analysing pus in bandages that were being discarded after use in surgery. He would have been amazed had he known that he was opening the door to one of the most glamorous areas of science that there has ever been. Miescher discovered a new chemical substance, which turned out to originate inside the nuclei of cells. Accordingly, he called it nuclein. Fifty years later, Phoebus Levine made inroads into its chemical structure, showing that Miescher's molecule was built from lots of copies of a basic unit, a nucleotide made from a sugar, a phosphate group and a base. He conjectured that the full molecule was made from a moderate number of copies of this nucleotide, attached to one another by the phosphate groups, and repeating the same pattern of bases over and over again.

When more had been discovered about his new molecule, it was named deoxyribonucleic acid,[1] which we all know by the acronym DNA. It was a gigantic molecule, and the techniques available then would never be able to reveal its structure – the atoms it contained, and how they were bonded to one another. But two decades later, the technique of X-ray diffraction was coming into use, and it proved to be just the ticket.

Light is an electromagnetic wave, and so are X-rays. When a wave encounters an obstacle, or passes through a series of closely spaced obstacles, it appears to bend. This effect is called diffraction. The exact mechanism depends on the mathematics of wave interference. The basic principles were discovered by the father-and-son team of William Lawrence Bragg and William Henry Bragg in

1913. The wavelength of X-rays is in the right range for them to be diffracted by the atoms in a crystal.

There are mathematical techniques for reconstructing the atomic structure of the crystal from the diffraction pattern that it produces. One of them is Bragg's law, which describes the diffraction pattern created by a series of equally spaced parallel layers of atoms, a particularly simple type of crystal lattice. The law can be used to deduce the spacing and orientation of such layers within a crystal. The mathematical concept that provides all of the fine details about how the atoms are arranged is the Fourier transform, introduced by the French mathematician Joseph Fourier in the early 1800s in a study of heat flow. Here the idea is to represent a periodic pattern in space or time as a superposition of regular waves of all possible wavelengths. Each such wave has an amplitude (how large the peaks and troughs of the wave are) and a phase (determining the precise positions of the peaks).

The main goal of X-ray diffraction is to find the electron density map of the crystal – that is, the way its electrons are distributed in space. From this, its atomic structure and the chemical bonds that hold the atoms together can be worked out. To do this, crystallographers observe the diffraction patterns produced by a beam of X-rays passing through a crystal. They repeat these observations with the crystal aligned at many different angles to the beam. From these measurements they deduce the amplitude of each component wave in the Fourier transform of the electron density. Finding the phase is much harder; one method is to add heavy metal atoms, such as mercury, to the crystal, and then compare the new diffraction pattern with the original one. The amplitudes and phases together determine the entire Fourier transform of the electron density, and a further 'inverse' Fourier transform converts this into the electron density itself. So, if you have an interesting molecule and can persuade it to crystallise, you can use X-ray diffraction to probe its atomic structure. As it happens, DNA can be made to crystallise, though not easily. In 1937 William Astbury used X-ray diffraction to confirm that the molecule has a regular structure, but he could not pin down what that structure was.

In the meantime, cell biologists had been figuring out what DNA *did*. There was certainly a lot of it about, so it ought to have some important function. In 1928 Frederick Griffith was studying the bacterium then called *Pneumococcus*, now *Streptococcus pneumoniae*, a major cause of pneumonia, meningitis and ear infections. The bacterium exists in two distinct forms. Type II-S is recognisable by its smooth surface, a capsule that protects it from the host's immune system, giving it time to kill the host. Type II-R has a rough surface – no capsule, hence no protection, so it succumbs to the host's immune system. Griffith injected live rough bacteria into mice, which survived. The same happened when he injected dead smooth bacteria. But when he injected a mixture of these two apparently harmless forms, the mice died.

This was surprising, but Griffith noticed something even more surprising. In the blood of the dead mice, he found *live* smooth bacteria. He deduced that *something* – he didn't know what, and in the time-honoured terminology of biology he gave it a vague name, the 'transforming principle' – must have passed from the dead smooth bacteria to the live rough bacteria. The explanation came in 1943, when Oswald Avery, Colin MacLeod and Maclyn McCarty showed that Griffith's 'transforming principle' was a molecule, DNA. The DNA of the dead smooth bacteria was somehow responsible for the existence of the protective capsule, and it had been taken up by the live rough ones – which promptly acquired their own capsules, and in effect turned into the smooth form. Presumably, the rough form does not have that particular *type* of DNA – though it does have its own DNA. The Avery–MacLeod–McCarty experiment strongly suggested that DNA was the long-sought molecular carrier of inheritance, and this was confirmed by Alfred Hershey and Martha Chase in 1952, when they showed that the genetic material of a virus known as the T2 phage is definitely DNA. The experiment also suggested that superficially identical molecules of DNA can be subtly different from one another.

The race was now on to determine the exact molecular structure of DNA. As so often happens in science, the key results came in a series of steps, not all of which were recognised to be significant when they were first discovered. It was already known from Levine's work that DNA was made from nucleotides, and each nucleotide was made from a sugar, a phosphate group and a base. It now transpired that there were four distinct bases: adenine, cytosine, guanine and thymine – all small, simple molecules (see Figure 15).

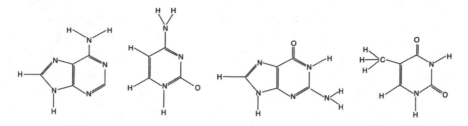

Fig 15 The four DNA bases. *Left to right*: Adenine, cytosine, guanine, thymine.

How did these four bases sit inside a complete DNA molecule? An important – but initially baffling – clue was found by the Austrian biochemist Erwin Chargaff, who had fled from the Nazis to the USA in 1935. Chargaff made careful studies of nucleic acids, including DNA, and in 1950 he pointed out a curious pattern. Table 4 shows some of his data for how frequently each base occurs in the DNA of various organisms, expressed as a percentage of the total number of bases.

Table 4   Some of Chargaff's data on the percentage of the four bases in the DNA of various organisms.

Organism	%A	%T	%G	%C
Human	29.3	30.0	20.7	20.0
Octopus	33.2	31.6	17.6	17.6
Chicken	28.0	28.4	22.0	21.6
Rat	28.6	28.4	21.4	20.5
Grasshopper	29.3	29.3	20.5	20.7
Sea urchin	32.8	32.1	17.7	17.3
Wheat	27.3	27.1	22.7	22.8

The numbers vary considerably from species to species – A occurs in 29.3% of humans, but 32.8% of sea urchins, for instance. However, there are some clear patterns. One is known as Chargaff parity rule 1. In each organism listed (and in many others), the percentages of A and T are almost equal, and the same goes for G and C. However, those of A/T can differ considerably from G/C. There is also a Chargaff parity rule 2, whose statement involves knowing (as we now do) that DNA consists of two intertwined strands. This states that the same equalities of percentages hold on each strand separately. In addition, Wacław Szybalski noticed that usually – though not always – the percentage of A/T is greater than that of G/C.

These three rules refer only to the overall percentages of the four bases in bulk DNA. They do not tell us – not directly – about the positioning of the bases within the molecule. We still do not know why Chargaff parity rule 2 and Szybalski's rule hold, but Chargaff parity rule 1 has a very simple explanation, which was one of the clues that led Crick and Watson to the famous double helix. They noticed that guanine and cytosine naturally join together using three hydrogen bonds, and similarly adenine and thymine join up using two hydrogen bonds (the dotted lines in Figure 16). Moreover, the two resulting pairs are chemically very similar – they have almost the same shape, the same size and the same potential for joining up with other molecules in the DNA structure.

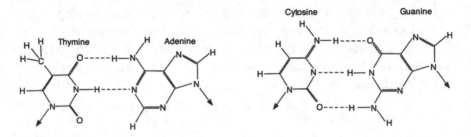

Fig 16   How bases pair up in DNA. The two joined pairs have almost identical shapes and sizes.

It was as though Chargaff had analysed a huge consignment of cutlery and crockery, and found that the percentages of knives and forks were the same, and similarly for cups and saucers. What might seem coincidence for the entire consignment made immediate sense if it were made up of packages each consisting either of paired knives and forks or paired cups and saucers. Then the percentages would match *exactly*, and they would match in every part of the consignment, not just in the overall totals. Similarly, if Crick and Watson were right, Chargaff parity rule 1 would be an immediate consequence. Not only were the percentages equal in bulk DNA, but the bases were arranged in matching pairs – like knives and forks.

This simple observation suggested that the DNA of organisms was made from these base pairs. When considered alongside other known features of the molecule, and an X-ray diffraction pattern obtained by Maurice Wilkins and Rosalind Franklin, it suggested a dazzlingly simple idea. DNA was a huge stack of base pairs, piled one on top of the other, held together by other parts of the molecule such as phosphate groups. The chemical forces between the atoms caused each successive base pair to be twisted by a fixed amount, relative to the one beneath. The base pairs were arranged like the slabs of a spiral staircase – or more accurately, two intertwined spiral staircases. The mathematical term for the shape is 'helix', so DNA was a double helix.

Watson's book of the same name told the story, warts and all, including his view of the difficulties encountered when trying to obtain access to Franklin's data, and how he and Crick dealt with them. By his own admission, this was not the pinnacle of scientific ethics, but was justified by the pressing need to sort out the structure before Nobel prize-winner Linus Pauling beat them to it. The story had a tragic ending. Franklin died of cancer, while Crick and Watson (and to a lesser extent Wilkins) got the glory.

When Crick and Watson published the double-helix structure of DNA in *Nature* in 1953, they pointed out that the occurrence of bases in specific pairs suggested an obvious way for DNA to be copied – which was necessary both for cell division and to transmit genetic information from parent to offspring. The point is that if you know one half of a base pair, you immediately know what the other half is. If one half is A, the other must be T; if one half is T, the other must be A. The same goes for G and C. So you could imagine some chemical process unzipping the two helical strands, pulling them apart, tacking on the missing half of each base pair and coiling the two resulting copies back into double helices.

If you think about the geometry, it is clear that this process can't be straightforward, and may not be a literal description. The strands will get tangled up for topological reasons – try separating the strands of a length of rope and you'll soon see why.

Biochemists now have strong evidence that the actual process involves several other molecules, types of enzyme, whose structure, intriguingly, is also coded in the organism's DNA.[2] Two of these enzymes, a helicase and a topoisomerase, unwind the double helix locally (Figure 17, see over). Then missing halves of the base pairs in the two separated strands are reconstituted, but not quite simultaneously. One strand leads and the other lags, probably because that makes room for the necessary molecular machinery to gain access and do the job. Another enzyme called DNA polymerase then fills in the matching pairs for the leading strand copy, while a second DNA polymerase does the same for the lagging strand. DNA polymerase makes its copies in short chains called Okazaki

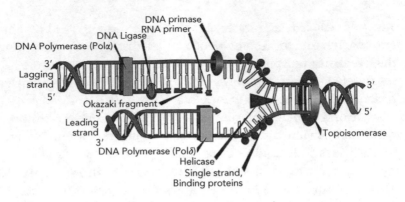

Fig 17   How to copy a DNA double helix. The entire process moves along the helix from left to right.

fragments. Then yet another enzyme worker, DNA ligase, joins the Okazaki fragments together.

You can see why Crick and Watson kept their speculations about replication brief, but also why they felt the need to say something, otherwise someone else would have made the same obvious suggestion and claimed the credit. And you can see why it took about fifty years to sort out how the trick was actually achieved.

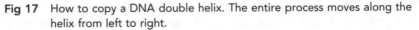

That was how DNA represented genetic 'information', and how the information was copied from parent to offspring. But what did the information *mean*?

The earliest proposal was that DNA is a recipe for proteins. Organisms are made from proteins – plus other things, but proteins are the most complex constituents, the most common and arguably the most important. Proteins are long chains of molecules known as amino acids, and twenty of these occur in living organisms. In an actual molecule the chain folds up in complex ways, but the key to making the chain is to specify the sequence of amino acids.

Soon after Crick and Watson's publication of the double-helix structure, the physicist George Gamow suggested that the most likely way for DNA sequences to specify amino acid sequences was by a three-letter code. His argument was a mathematical thought experiment. Using the four bases as letters, you can form 4 one-letter words (A, C, G, T), $4 \times 4 = 16$ two-letter words (AA, AC, ..., TT), and $4 \times 4 \times 4 = 64$ three-letter words (AAA, AAC, ..., TTT). Using one *or* two letters provided a total of 20 possible words – but how could the chemistry tell that AA is a two-letter word rather than two separate A's? It made more sense to use words of a fixed length, which had to be at least three since there were too few two-letter words to specify 20 amino acids. With 64 three-letter words to play with there is plenty of wiggle room, so it seemed highly inefficient to use even more letters.

A long series of brilliant experiments proved Gamow right, and led to what we now call the genetic code. It's worth looking at this in detail, because it displays a puzzling mixture of pattern and irregularity.

Table 5   The 20 amino acids involved in the genetic code, and which triplet of DNA bases codes for which amino acid.

Alanine (Ala), arginine (Arg), asparagine (Asn), aspartic acid (Asp), cysteine (Cys), glutamic acid (Glu), glutamine (Gln), glycine (Gly), histidine (His), isoleucine (Ile), leucine (Leu), lysine (Lys), methionine (Met), phenylalanine (Phe), proline (Pro), serine (Ser), threonine (Thr), tryptophan (Trp), tyrosine (Tyr), valine (Val)

TTT	Phe	TTC	Phe	TTA	Leu	TTG	Leu
TCT	Ser	TCC	Ser	TCA	Ser	TCG	Ser
TAT	Tyr	TAC	Tyr	TAA	STOP	TAG	STOP
TGT	Cys	TGC	Cys	TGA	STOP	TGG	Trp
CTT	Leu	CTC	Leu	CTA	Leu	CTG	Leu
CCT	Pro	CCC	Pro	CCA	Pro	CCG	Pro
CAT	His	CAC	His	CAA	Gln	CAG	Gln
CGT	Arg	CGC	Arg	CGA	Arg	CGG	Arg
ATT	Ile	ATC	Ile	ATA	Ile	ATG	Met/START
ACT	Thr	ACC	Thr	ACA	Thr	ACG	Thr
AAT	Asn	AAC	Asn	AAA	Lys	AAG	Lys
AGT	Ser	AGC	Ser	AGA	Arg	AGG	Arg
GTT	Val	GTC	Val	GTA	Val	GTG	Val
GCT	Ala	GCC	Ala	GCA	Ala	GCG	Ala
GAT	Asp	GAC	Asp	GAA	Glu	GAG	Glu
GGT	Gly	GGC	Gly	GGA	Gly	GGG	Gly

Table 5 lists the 20 amino acids involved and which triplet of DNA bases codes for which amino acid. To bring out the structure,

I've split the triplets into four blocks, according to the first letter; then within each block, the second letter corresponds to the row and the third letter to the column. The triplets TGA, TAA, TAG and ATG are exceptional. The first three of these do not code for an amino acid, but stop the conversion of triplets into amino acids. When the fourth triplet ATG is at the beginning of a gene it causes this process to start, but otherwise it codes for methionine.

We speak of 'the' genetic code because exactly the same code applies to the nuclear DNA sequences of virtually every known microorganism, plant and animal. There are a very small number of exceptions, almost all of which assign an amino acid to one of the three STOP triplets. Also, mitochondria contain their own DNA, and the genetic code for mitochondrial DNA has some small but important differences.[3] A very similar code applies to DNA's sister molecule RNA (ribonucleic acid), which among other things plays a key role in turning DNA code into proteins, a process known as transcription. The code in RNA has exactly the same structure, but with thymine (T) replaced by a similar molecule, uracil (U). You will often see U used in place of T in sequence data.

The genetic code contains both intriguing hints of order and baffling irregularities. There are more triplets than amino acids (even counting STOP and START), so some amino acids must be specified by several triplets. The most common number is four triplets to an amino acid: for example TCT, TCC, TCA and TCG all code for serine (the top row of triplets in the table). Here the third letter tells us nothing new, so it is redundant as far as the code goes. This feature is common: eight of the sixteen rows in the table correspond to precisely one amino acid. The third letter could not be removed entirely in these cases, because the current chemical machinery works with triplets. But this pattern suggests that maybe there was once a simpler code, using pairs of bases, and TC indicated serine. Later, the code itself evolved.

However, phenylalanine corresponds to TTT and TTC, but TTA and TTG correspond to leucine. So do four triplets that begin with CT. So here two triplets code for phenylalanine, but a whopping *six* for leucine. It is as if the genetic code itself also evolved – but how could it, while still 'working'?

There are mathematical issues here, to do with changes to codes and why they occur, as well as biological questions. Any major changes in the code must have happened very early on, when life was first appearing on Earth, because once life gets going, changes at such a fundamental level become increasingly difficult, and we ought to see remnants of discarded codes somewhere in the living world, albeit in some obscure organism. As I've said, there are a few exceptions to the standard genetic code, but those look like recent slight changes, not ancient major ones.

The 'frozen accident theory' takes the view that the genetic code could easily have been very different. A different code would have worked fine, but when life began to diversify, whichever code was in use could no longer survive major changes. A code that initially arose by accident was frozen in place once variants couldn't compete with the established code.

However, there are tantalising hints that the current code might be more 'natural' than most alternatives. There are biochemical affinities between particular amino acids and particular triplets, so the whole set-up might have been predisposed towards some relatively minor variant of the code we find today.

Attempts are being made to trace the likely evolution of the early code, before it became locked in, based on symmetry principles and a host of other speculative uses of mathematics and physics, to try to reconstruct whatever it was that happened about 3.8 billion years ago. Finding definitive evidence, when DNA data for organisms more than a few tens of thousands of years old are unobtainable, will be tricky. So this one will run and run.

. . . .

DNA provided an unexpected source of independent evidence in favour of evolution.

Before Crick and Watson discovered the structure of DNA, taxonomists had developed Linnaeus's classification scheme into an extensive description of the 'Tree of Life', the presumed evolutionary ancestry of today's organisms, by comparing the anatomical features and behaviour of existing creatures. This complex sequence of divergent species constituted a massive collection of predictions based on evolutionary theory. What was needed was an independent way to test those predictions – and DNA sequencing does exactly that. When suitable techniques became available, it turned out that organisms thought to be related by evolution generally have similar DNA sequences.

In some cases it is even possible to determine a specific genetic change that distinguishes two species. Lacewings are elegant, delicate flying insects which occupy some 85 genera and about 1,500 species. In North America the two most common species are *Chrysopa carnea* and *C. downesi*. The first is light green in spring and summer but brown in autumn, whereas the second is permanently dark green. The first lives in grasslands and deciduous trees; the second lives on conifers. The first breeds twice, once in the winter and once in summer; the second breeds in the spring.

In the wild the two types do not interbreed, so they satisfy the usual definition of distinct species. However, the genetic differences between them are small and very specific, and centre on three genes. One controls colour, and the other two control the breeding time via the insect's response to the length of daylight (a common mechanism found in insects to ensure that they breed at a suitable time of year). These genetic differences were initially inferred from laboratory experiments in which the two species were brought together and given the opportunity to interbreed.

Natural selection suggests a simple explanation of these findings. A dark green insect is less visible on a dark green conifer than a light green one is. So when dark mutants appeared, natural selection kicked in: dark green lacewings that lived on conifers were better protected against predators than light green ones, and conversely light green ones living on light green grass were better protected than dark green ones. Once the two incipient species came to occupy separate habitats, they ceased to interbreed – not because this was impossible in principle, but because they didn't meet up very often. The accompanying changes to the breeding time reinforced this 'reproductive isolation'.

Detailed analysis of a truly gigantic number of DNA sequences leads to a Tree of Life that is very similar to the one already developed on purely taxonomic grounds. The correspondence isn't perfect – it would be very suspicious if it were – but it is strikingly good. So DNA mutations joined the story, confirming most of the evolutionary sequences that taxonomists had inferred from phenotypes, and providing clear evidence of the underlying genetic mutations that enabled phenotypic changes.

A fossil rabbit in the wrong geological stratum would disprove evolution. So would a modern rabbit with the wrong DNA.

. . . .

The supremacy of the gene has become entrenched in our collective consciousness to such an extent that news media often talk of 'genetic science' when no genes or DNA are involved at all. The vivid, flawed, but by no means nonsensical image of the 'selfish gene' introduced in Richard Dawkins' book of the same title has captured public imagination.[4] The main motivation for this image is the existence of so-called junk DNA,[5] which gets replicated along with the important parts of the genome when an organism reproduces. From this, Dawkins argues that the survival of a segment of DNA is the sole criterion for its successful replication, whence DNA is 'selfish'.[6]

This is of course the old 'a chicken is just an egg's way to breed a new egg' line of thinking, dressed up in fancy hi-tech. It is equally possible to give an account of genetics and evolution from the point of view of an organism, leading to what biologist Jack Cohen and I call the 'slavish gene', obsessed with activities that do not damage the organism's survival chances.[7] The selfish gene metaphor is not wrong – in fact, it is intellectually defensible as a debating point. But it diverts our attention rather than adding to our understanding. The relation between genes and organisms is a feedback loop: genes affect organisms via development; organisms affect genes (in the next generation) via natural selection. It is a fallacy to attribute the dynamics of this loop to just one of its components. It's like saying that wage increases cause inflation, but forgetting that price increases fuel demands for higher wages.

The image of the selfish gene has also inspired a rather naive kind of genetic determinism, in which the only things that matter about human beings are their genes. This vision of the gene as absolute dictator of form and behaviour lies not far beneath the surface of today's biotechnology: so-called 'genetic engineering'. By cutting and splicing DNA molecules, it is possible to insert new genes into an organism or to delete or otherwise modify existing ones. The results are sometimes beneficial: pest resistance in agricultural plants, for instance. But even this can have undesirable side effects.

Such technology is controversial, especially when it comes to genetically modified food. There are good reasons for this controversy, and both sides have put forward some compelling arguments. My own feeling is that we know enough about the mathematics of complex systems to be very wary of simplistic models of how genes act, and this feeling is reinforced by the huge amount that we don't know about genes. I wouldn't trust a computer whose software had been hacked by a bright ten-year-old, even if the result gave me a really nice screen saver. I would worry whether anything had accidentally been damaged when the program was hacked. I would be more likely to trust a professional programmer who really understood the computer's operating system. But right now, genetic engineers are really just clever hackers, and no one has much idea of how the genetic 'operating system' really works.

With huge amounts of money at stake, the discussion has become polarised. Opponents of genetic modification are often branded as 'hysterical', even when they make their criticisms in a moderate and well-reasoned manner. Biotechnology companies are accused of taking huge risks for profit even though some of their motives are more benign. Some people deny that any risks exist; others exaggerate them. Underneath these arguments is a serious scientific problem which deserves more attention. To put it bluntly: our current understanding of genetics is completely inadequate for assessing the likely benefits, costs or potential dangers of genetic modification.

This may seem a strong claim, but specialists in genetics are often blinded by their own expertise. Knowing far more about genes and their modification than the opposition, they fall into the trap of thinking that they know everything. However, the entire history of genetics shows, that at every stage of its development, whenever new data became available most of the previous confidently held theories topple in ruins.

Only a few decades ago, each gene was thought to occupy a single connected segment of the genome, and its location was fixed. Barbara McClintock, a geneticist at Cornell University, made a series of studies of maize and deduced that genes can be switched on or off, and that they can sometimes move. For years her ideas about 'jumping genes' were derided, but she was right. In 1983 she won a Nobel prize for discovering what are now called transposons – mobile genetic elements.

Before the human genome was sequenced – a story I'll tell in the next chapter – the conventional wisdom was that one gene makes one protein, and since humans have 100,000 proteins, they must have 100,000 genes. This was pretty much considered to be a fact. But when the sequence was obtained, the number of genes was only a quarter of that. This unexpected discovery drove home a message that biologists already knew, but had not fully taken on board: genes can be chopped up and reassembled when proteins are being made. On average, each human gene makes four proteins, not one, by exploiting this process.

Each overturning of the conventional wisdom can be viewed in a positive light: human knowledge is thereby advanced, and we gain insight into life's subtleties. But there is a negative aspect as well: the overwhelming confidence that the system was thoroughly understood, and that no big surprises were going to happen, right up to the moment when it all fell to bits. This does not set a convincing precedent.

Genetic modification has huge potential, but there is a danger of this being squandered by prematurely bringing experimental organisms to market. The commercial use of genetically modified food plants has already led to unexpected adverse effects, and hardly any of the plants have lived up to the early hype. Most were quickly withdrawn. Some that initially appeared to be successes are running into trouble. There is a tendency among biotechnology companies to focus on food safety ('our grain is perfectly safe'), where they feel comfortable. They tend to ignore potential undesirable effects on the environment, especially delayed-action effects, our knowledge of which is pitiful – mainly because we don't understand ecosystems well enough. No amount of genetic expertise will improve that.

However, safety is also a significant concern. The argument that genetic modification merely does quickly and directly what conventional plant breeding does slowly and indirectly is nonsense. Conventional breeding mimics nature by forming new combinations using existing genes, through the operation of the plant's normal genetic machinery. Genetic modification fires alien DNA randomly into the genome, allowing it to lodge wherever it falls. But an organism's genome is not merely a list of bases. It is a highly complex dynamical system. It is naive to imagine that making crude changes here and there will have only the obvious, expected effects.

Imagine taking a gene whose effect, in its normal location in its normal organism, is to make a protein that has no adverse effect on humans – which is what 'safe to eat' basically means. Does that guarantee that it will be equally 'safe' when introduced into a new organism? On the contrary: it could potentially wreak havoc, because we often don't know where the new segment of DNA will lodge itself – and even if we do, genes can move. The new gene might not make the desired protein and nothing else. It might not make it at all. It could end up inside another gene, interfering with that gene's function. This function might be making a protein: if so, either the wrong protein gets made, with potential hazards, or it doesn't get made, with knock-on effects for the whole plant. Worse, the newcomer might end up inside a regulatory gene, and the entire network of gene interactions could go haywire.

None of this is particularly likely, but it is possible. Organisms reproduce, so any disaster can propagate and grow. We have the perennial problem of an event that is very unlikely, but could do enormous damage if it happens, with the added feature that it can reproduce.

In the rush to market, experiments have been carried out on a large scale in the natural environment, when controlled laboratory testing would have been far more effective and informative. The British Government sanctioned large-scale planting of genetically modified plants in order to test whether their pollen spread only a few metres (as expected) and to make sure that the new gene would not be spontaneously incorporated into other species of plant (ditto). It turned out that the pollen spread for miles, and the new genes could transfer without difficulty to other plants. Effects like this could, for example, create pesticide-resistant strains of weeds. By the time the experiment had revealed that the conventional wisdom was wrong, there was no way to get the pollen, or its genes, back. Simple laboratory tests – such as painting pollen onto plants directly – could have established the same facts more cheaply, without releasing anything into the environment. It was a bit like testing a new fireproofing chemical by spraying it on a city and setting the place alight, with the added twist that the 'fire' might spread indefinitely if, contrary to expectations, it took hold.

It is all too easy to imagine that the genome is a calm and orderly place, a repository of information that can be cut and pasted from one organism to another, only ever performing 'the function' that geneticists expect it to perform. But it's not – it's a hotbed of dynamic interactions, of which we understand only the tiniest part. Genes have many functions; moreover, nature can invent new ones. They do not bear a label 'USE ONLY TO MAKE PROTEIN X'.

Continuing to do research on genetic modification, and occasionally using successfully modified organisms for specific purposes such as the production of expensive drugs, make good sense. Helping developing countries to produce more food is a worthy aim, but it is sometimes used as an excuse for an alternative agenda, or as a convenient way to demonise opponents. There is little doubt that the technology needs better regulation: I find it bizarre that standard food safety tests are not required, on the grounds that the plants have not been changed in any significant way, but that the innovations are so great that they deserve patent protection, contrary to the long-standing view that naturally occurring objects and substances cannot be patented. Either it's new, and needs testing like anything else, or it's not, and should not be patentable. It is also disturbing, in an age when commercial sponsors blazon their logos across athletes' shirts and television screens, that the biotechnology industry has fought a lengthy political campaign to prevent any mention of their product being placed on food. The reason is clear enough: to avoid any danger of a consumer boycott. But consumers are effectively being force-fed products that they may not want, and whose presence is being concealed.

Our current understanding of genetics and ecology is inadequate when it comes to the widespread use of genetically modified organisms in the natural environment or agriculture. Why take the risk of distributing this material, when the likely gains for most of us – as opposed to short-term profits for biotechnology companies – are tiny or non-existent?

. . . .

It was once thought that an organism's DNA contains all of the information required to determine its form and behaviour. We now know that this is not the case. The genome is of course very influential, but several other factors can affect the developing organism. Collectively, they are known as epigenetic features. The word epigenetics means 'above genetics'. It refers to changes in phenotype or gene expression that can be passed on to the next generation, but do not reside in DNA.

Among the first epigenetic processes to be discovered was DNA methylation. Here a region of DNA acquires a few extra atoms, a methyl group. This causes a cytosine base to change into a closely related molecule, 5-methylcytosine. This modified form of cytosine still pairs up with guanine in the DNA double helix, but it tends to 'switch off' that region of the genome, with the result that the proteins that are being encoded are produced in smaller quantities.

Another is RNA interference. This remarkable phenomenon is enormously important, yet it was not discovered through a major research programme: several biologists discovered the effect independently. One of them was Richard Jorgensen. In 1990 his research team was working on petunias, hoping to breed new varieties with brighter colours. They started by trying an obvious piece of genetic modification: engineering extra copies of the pigment-producing gene into the petunia genome. Obviously, more enzyme would produce more pigment.

But it didn't. It didn't produce less pigment, either. Instead, it made the petunia stripy.

Eventually, it transpired that some RNA sequences can switch a gene off, and that stops it making protein. The stripes appeared because the pigment genes were switched on in some cells and off in others. This 'RNA interference' turns out to be very common. It opens up the prospect of deliberately switching genes on or off, which would be important in genetic engineering. More fundamentally, it changes how biologists view the activity of the gene.

The orthodox picture, as I've said, was that each gene makes one protein, and each protein has one function in the organism. For instance, the haemoglobin gene makes haemoglobin, and haemoglobin carries oxygen in the blood and releases it where it is needed. So a specific sequence in an organism's DNA can be translated directly into a feature of the organism. But as the geneticist John Mattick wrote in *Scientific American*[8]:

> Proteins do play a role in the regulation of eukaryotic gene expression, yet a hidden, parallel regulatory system consisting of RNA that acts directly on DNA, RNAs and proteins is also at work. This overlooked RNA signalling network may be what

allows humans, for example, to achieve structural complexity
far beyond anything seen in the unicellular world.

Some epigenetic effects involve DNA, but in a different organism.
In mammals, for instance, the early stages of an egg's development
are controlled, not by the egg's DNA, but by that of the mother.
This actually makes a lot of sense, because it lets a fully functioning
organism kick-start the growth of the next generation. But it means
that a key stage in the growth of, say, a cow is not controlled by
that cow's DNA. It is controlled by another cow's DNA.

Even more broadly, some things can be conveyed from parent
to child through cultural interactions rather than genetic ones. This
effect is very common in humans – we acquire our language from
our culture, our religious beliefs or lack of them, and many other
things that make us human. But behaviour is acquired by similar
cultural interactions in rats, dogs and many other animals.

. . . .

When the genetic code was discovered, DNA was seen as a kind of
blueprint. Once you possess an engineering blueprint of, say, an
aircraft, then any competent engineer will be able to tell you how
to make it. Once you possess 'the information' to make an animal,
you can make that animal. And if you can make it, you must know
everything there is to know about it.

It stands to reason.

Well, no, it doesn't. Put baldly like that, it sounds like an
obvious exaggeration, a pun in which different meanings of the
word 'information' are confused. It doesn't even work for
engineering. You need a lot more than just 'the blueprint' to build
an aircraft. You need to know all the engineering techniques that
are implicit in the blueprint. You need to know how to make the
components, how to choose and obtain suitable materials, and you
need the right tools.

It works even less well for biology, where the analogous
'techniques' are implicit in the way the organisms themselves work.
You can't make a baby tiger from tiger DNA. You need a mother
tiger too – or at the very least, you need to know how a mother
tiger does the job. And even if this were implicit in her DNA (which
it's not, because of epigenetic effects), you would have to make the
implicit explicit. Despite that, the vision of DNA as King led to

enormous progress in biology. It, and the associated discoveries about biological molecules, are a major reason why today's doctors can actually cure many diseases. For previous generations, this was essentially impossible.

The genetic sequences encoded in DNA are a big part of 'the secret of life'. If you're not aware of the role of DNA, if you don't know what the sequence looks like, you're missing a gigantic part of the picture. It's like trying to figure out how modern society works when you don't know about telephones.

But DNA is not the *only* secret.

Figuring that out took much longer, and was more discouraging. When you've made such a huge breakthrough, one that takes you so far in comparison with everything that has gone before, it's disappointing to discover that unlocking one boxful of secrets and raising the lid does not do a Pandora and reveal all manner of biting insects and vile creatures, but just reveals ... another locked box inside.

8 The Book of Life

· ·

In 1990 the world's geneticists embarked on the most ambitious programme of biological research the world had ever seen, which many compared in its scale to the Kennedy-era project to land a man on the Moon. Biologists were aiming to join the ranks of big science, previously occupied mainly by particle physicists, nuclear physicists and astronomers, where governments were willing to spend billions of dollars rather than mere millions. This financial aim was explicit, but the scientific objective was impeccable and important: to sequence the human genome – that is, to obtain the complete sequence of DNA bases in a typical human. It was known that there are about three billion of these, so the task would be difficult, expensive, but feasible. Just right for big science.

The project emerged from a series of workshops supported by the US Department of Energy, beginning in 1984 and leading to a report in 1987.[1] This set the goal of sequencing the human genome, pointing out that this objective was 'as necessary to the continuing progress of medicine and other health sciences as knowledge of human anatomy has been for the present state of medicine'. In the popular media, biologists were in search of the Book of Life.

In 1990 the Department of Energy and the National Institutes of Health announced a $3 billion project – one dollar per base pair. Several other countries joined the USA to create a consortium: Japan, the United Kingdom, Germany, France, China and India. At that time, finding even short DNA sequences was time-consuming and laborious, and it was estimated that the project would take fifteen years. This estimate was not far out, despite huge improvements in sequencing technology. However, a latecomer to the game demonstrated that the whole project could have been completed in about three years, for one-tenth of the cost, by putting more effort into thinking and less into complicated biochemistry. In 1998 Craig Venter, a researcher at the National Institutes of Health, founded his own company, Celera Genomics, and set out to derive the entire sequence independently using $300 million contributed by private investors.

The publicly funded Human Genome Project (HGP) published its new data on a daily basis, and announced that all its results would be freely available. Celera published its data annually, and announced its intention to patent some of it – a few hundred genes. In the event, Celera initiated preliminary patents on 6,500 genes or partial genes. These intellectual property rights were what, with luck, would repay investors. A corollary was that Celera's data would not be free for all researchers to use; the company's initial agreement to share data with HGP came to bits when Celera declined to lodge its data in the publicly accessible GenBank database. But Celera used HGP's data as part of its own effort – well, it was public.

To protect the free release of vital scientific data, HGP took steps to publish its data first, which would (subject to legal wranglings) constitute 'prior art' and invalidate Celera's patents. In the event, HGP published the 'final' sequence a few days before Celera. By then, President Bill Clinton had already stated that he would not permit the genome sequence to be patented, and Celera's market value plummeted. The NASDAQ stock exchange, overloaded with biotechnology companies, lost tens of billions of dollars.[2]

In 2000 Bill Clinton and Tony Blair announced to the world that a 'draft genome' had been obtained. The next year, both HGP and Celera published drafts which were about 80% complete. An 'essentially complete' genome was announced by both groups in 2003, though there were disagreements about what this phrase meant, but improvements in the period up to 2005 led to a sequence that was about 92% complete. The main stages of the programme were to sequence complete chromosomes – recall that we humans have 23 pairs of these. The sequence for the final human chromosome was published in *Nature* in 2006.

By 2010 most of the gaps in the sequence had been filled, although a significant number remain. So do numerous errors. A dedicated group of geneticists has undertaken the task of filling the gaps, and eliminating the errors in supposedly 'known' regions. This is scientifically essential, but they will get little credit for it because the exciting frontier has moved on. Their devotion to science is admirable.

* * * * * * * * * * * *

How do you 'read' the sequence of such a huge molecule as DNA? Not by starting at one end and proceeding along it to the other.[3] Current sequencing techniques do not work on gigantic molecules: they typically use stretches of 300–1,000 DNA bases. The way round this restriction is obvious, though not its implementation: break the molecule into short fragments, sequence them, then stick them all back together in the right order.

The first effective sequencing method goes back to Allan Maxam and Walter Gilbert in 1976. Their idea was to change the structure of the DNA molecule at specific bases, and add a radioactive 'label' at one end of each fragment. Four different chemical processes then targeted the four types of base. It would have been neat and tidy if these processes could cut the strand at A, C, G and T respectively, but the chemistry didn't work out like that. Instead, two processes created cuts at specific bases, C and G, while the others had a little ambiguity: they created cuts at either of two distinct bases: A or G, and C or T. However, if you knew the 'A or G' data and the G data, you could deduce which of the 'A or G' cuts were A and which were G, and similarly for the 'C or T' cuts. Knowing which type of base is located at the cut, and using the radioactive label to sort the fragments into order by letting them diffuse through a sheet of gel, you could deduce the sequence of bases. This method is called gel electrophoresis, because it passes an electric current through the gel to make the molecules diffuse.

The next advance was the chain-terminator method, also called the Sanger method after its inventor, Frederick Sanger. This procedure also creates fragments of varying lengths, which are similarly diffused through a gel to sort them into order. The cunning step is to attach fluorescent dyes to the molecular labels – green for A, blue for C, yellow for G, red for T – and to read these automatically using optical methods.

This technique works well for relatively short strands of DNA, up to about a thousand bases, but it becomes inaccurate for longer strands. Numerous technical variations on the chain-terminator method have been devised to streamline the process and speed it up. Statistical methods have been developed to improve accuracy, in cases where the patch of fluorescent dye is a bit faint or fuzzy. Automated DNA sequencers can handle 384 DNA samples in a single run, and can carry out about one run per hour. So in a day, you can sequence about 9,000 strands – a little under 10 million bases per sequencer per day. As experience grows and demand for new sequences increases, these numbers are rising rapidly.

. . . .

At this stage there is a trade-off between two aspects of the problem. Creating the break points and sequencing the resulting fragments are a matter of biochemistry. The cleverer you are when you break the molecule up, the easier it will be to reassemble the pieces. If you cut the chain at 'known' locations, and keep track of which pieces are adjacent to them, then reassembly is in principle straightforward. It's like marking the corresponding ends of the fragments with labels that match.

If convenient break points always existed, this method would be very effective. But often they don't, and then an alternative is required. Both the HGP and Celera used the 'shotgun' method. This breaks the strand into random fragments. Each fragment is sequenced, then they are fitted together by mathematical techniques, implemented on fast computers. The method works because the random fragments sometimes overlap, where the same bit of DNA has been cut in two different places. Those overlaps tell you how the pieces fit together.

In oversimplified terms, it works like this. Suppose you have two pieces

CCTTGCCAAA and TGTGTGAACC

and you know that they abut, but not in which order. Then you have to decide whether the correct join is

CCTTGCCAAATGTGTGAACC

or

TGTGTGAACCCCTTGCCAAA

If you have an overlapping fragment that reads

GAACCCCTTG

then it fits the second possibility

TGTGTGAACCCCTTGCCAAA

but not the first. In practice, you have a lot of these fragments, and a variety of information about possible ways to reorder them. And the fragments are much longer – but that helps more than it hinders, because the overlaps can be larger, hence less ambiguous.

Again, there are lots of different ways to carry out this strategy for sequencing long DNA strands. All of them rely heavily on computers to do the mathematical calculations and to handle the large quantities of data, but a lot of mathematical thought has to go into sorting out what the computers are instructed to do. A simple example of such methods is the so-called greedy algorithm. Given a collection of fragments, many of them overlapping, first find the pair of fragments with the biggest overlap. Merge them into a single chain and replace them by this new chain. Now repeat. Eventually, many of the fragments will be merged into a single chain; with enough overlaps, all of them. This method does not always lead to the shortest chain that is consistent with all the fragments, and it may not produce the correct assembly. It is also computationally inefficient because at each stage you have to calculate all the overlap sizes for all the pairs of fragments.

As a simple example with much smaller numbers, suppose the fragments are

TTAAGCGC   CCCCTTAA   GCTTTAAA   TCCCCCCA

The biggest overlap occurs with CCCCTTAA and TTAAGCGC, which therefore merge to give CCCCTTAAGCGC, and that replaces the first two sequences on this list. The biggest overlap among what's left is with CCCCTTAAGCGC and GCTTTAAA, which merge to give CCCCTTAAGCGCTTTAAA. The fourth sequence on the list doesn't overlap this, so it has to be left unconnected until further data are obtained.

The HGP put most of its money on the biochemical step, and first broke up the genome into sequences of about 150,000 bases by cutting it at specific locations. This involves a lot of effort finding enzymes that make suitable cuts, and sometimes these prove elusive. Then each of these fragments was sequenced by the shotgun method. Celera put all of its money on the mathematical step, and applied the shotgun to the entire human genome. Then it used a large number of sequencing machines to find the DNA sequences of the fragments, and assembled them by computer.

You can either use clever chemistry to simplify the maths, or use clever maths to simplify the chemistry. HGP took the first approach, Celera the second. The second turned out to be cheaper and faster, mainly thanks to the enormous power of modern computers and the development of slick mathematical methods. The wisdom of this choice was not entirely obvious at first, because Celera used data from the HGP in its assembly process. But as more and more genomes have been sequenced, it has become clear that whole-genome shotgun is the way to go, at least until something better comes along.

Today, sequencing genomes has become almost routine. Scarcely a week passes without an announcement that a new organism has been sequenced – over 180 species to date. Most are bacteria, but they also include the mosquito responsible for infecting humans with malaria, the honeybee, the dog, the chicken, the mouse, the chimpanzee, the rat and the Japanese spotted green pufferfish. As I write, the latest is a sponge, whose sequence may shed light on the origin of eukaryotes.

In *Jurassic Park*, dinosaurs were brought back to life by sequencing their DNA, extracted from blood that had been ingested by blood-sucking flies which were then preserved in amber. This fictional technique doesn't work in reality, because ancient DNA degrades too fast, but over the past few years something similar has been done with DNA that is tens of thousands of years old. In particular, we now have a growing understanding of the Neanderthal genome. Neanderthals, of course, were a rather robust form of hominid that coexisted with early modern humans, between about 130,000 and 30,000 years ago. Until recently, some taxonomists have considered them to be a separate species, *Homo neanderthalensis*, while others have classified them as a subspecies, *H. sapiens neanderthalensis*, of *H. Sapiens.* It is now known that about 4% of people alive today have some DNA sequences derived from the Neanderthal genome, transmitted via a Neanderthal male and a modern-human female. So DNA supports the subspecies classification.

. . . .

The word 'gene' is bandied about as if everyone knows what it means. Genes make you what you are. They explain everything about you. Genes make you fat, they make you homosexual, they cause diseases, they control your destiny.

Genes are magic. Genes perform miracles.

It is worth distinguishing two uses of the word 'gene'. One is very limited: a gene is a portion of the genome (not necessarily in one connected piece) that codes for one or more proteins. Not so long ago the conventional wisdom would have deleted 'or more', but the Human Genome Project revealed that although we have 100,000 different proteins in our bodies, they are specified by only 25,000 genes. Genes often come in several pieces, and the amino acid sequences that these pieces specify can be spliced together in many different ways. So the same gene can, and often does, code for several different proteins.

The second usage of 'gene' is extraordinarily broad. It arises from the activities of neo-Darwinists, who reinterpreted Darwinian evolution in terms of DNA. This approach has enormous scientific value, but some of the interpretations associated with it are questionable; and unfortunately these interpretations have become common currency, while only specialists understand the underlying science. In his elegant masterpiece *The Blind Watchmaker*, Richard Dawkins defined the phrase 'a gene for X' to mean 'any kind of genetic variation that affects X'. Here X is any feature of an organism; Dawkins' example is 'tying shoelaces'.

This definition is just about defensible, though it runs into trouble when X is 'having a blue-eyed mother'. In practice, 'affects X' is interpreted as 'changes to the gene correlate with changes to X', because cause and effect are often hard to establish. The children of blue-eyed mothers do indeed exhibit genetic variation that correlates with having a blue-eyed mother, so in that sense they 'have a gene for having a blue-eyed mother'. But in the stricter sense the gene that matters is actually in the mother; the children show genetic variation because they sometimes inherit that gene.

Even ignoring such examples, the second, abstract definition of 'gene' can cause problems if it is confused with the first, concrete definition. Predictably, that's exactly what has happened. Many people now assume that our behavioural quirks and predisposition to various diseases can be traced to specific DNA sequences in our genetic make-up. Newspaper reports of geneticists finding 'the gene for' something or other can lead us to believe that the something or other in question is somehow written into our (some of us's) genes. There are genes – we are told – for blue eyes, cystic fibrosis, obesity, novelty-seeking, susceptibility to heroin addiction, dyslexia, schizophrenia and emotional sensitivity. Analyses of identical twins separated at birth suggest that your genes may even determine what kind of person you marry and what make of car you buy.

Somewhere in my genes, apparently, it says 'Toyota'. I find this very curious, because according to my birth certificate my genes existed in 1945, but Toyota made no significant exports to the UK until the 1970s.

Alleged 'genes for' schizophrenia, alcoholism and aggression have been announced, to a flourish of trumpets, and then quietly withdrawn when subsequent evidence fails to back up the initial assertion. The location of a gene for breast cancer has been claimed several times, not always correctly. Biotechnology companies have fought in court over patent rights to genes that are thought to increase the risk of contracting various diseases.

In 1999 the *Guardian* newspaper printed an article with the headline ' "Gay gene" theory fails blood test.'[4] This story began in 1993 when a segment of human chromosome known as Xq28, inherited from the mother, was implicated in male homosexuality. The initial evidence came from a study of gay male twins and brothers carried out by Dean Hamer and others, which concluded that gay men tend to have more gay relatives, on the maternal side, than heterosexual men do.[5] Later, various researchers found that in 40 pairs of gay brothers, the genetic similarities in the Xq28 region were significantly greater than chance. This finding created a global media sensation, and the 'gay gene' seemed to have been given sound scientific basis, even though no scientist ever claimed to have pinned anything down to a single gene.

The fateful chromosome segment Xq28 played a central role in Hamer's book *Living with Our Genes*, but even before it appeared, serious doubts were surfacing. In particular, other researchers couldn't replicate Hamer's results. In 1999 the journal *Science* carried an article by George Rice and colleagues, who examined blood from 52 pairs of brothers and attempted to confirm the link between Xq28 and homosexuality. They reported: 'Our data do not support the presence of a gene of large effect influencing sexual orientation at position Xq28.'[6]

This negative conclusion remains in force today. In fact, the role of individual genes in determining large-scale human characters – those we encounter on a human level – seems to be very small. Leaving aside a few direct connections, such as hair and eye colour, the link between any specific gene and a human-level character is virtually non-existent. As evidence, consider height. There is little doubt that people's genes play a major role in determining their height: tall parents tend to have tall children. So it is no surprise that, to date, height is the character that has been found to be most closely correlated with the presence or absence of a single gene (again excepting hair colour and the like). What is a surprise, however, is the extent to which this particular gene affects height. It accounts for an astonishing ... *two per cent* of the variation in human height.

And that's the *biggest* correlation between a single gene and a human character.

. . . .

How can two competent studies, both using similar methods, lead to such contradictory results? I'm not suggesting that the scientists concerned acted in any way improperly. But there is a mechanism that can easily lead to these kinds of contradictory outcomes, even though the experiments have been performed honestly and competently. It comes from a subtle misinterpretation of statistics.

Statistical methods are used to assess correlations between two data sets. For instance, heart disease and obesity in humans tend to be associated. The degree of correlation can be calculated mathematically; its statistical significance is a measure of how likely it is for such a correlation to have arisen by pure chance. If, for instance, that level of correlation occurs 1% of the time in randomly chosen data sets, then the correlation is said to be significant at the 99% level.

The widespread availability of computer software has rendered virtually effortless a procedure that not long ago required days of work on a desktop calculating device – what we might call the 'scattershot' approach to finding significant correlations between genes and characters. Suppose you start out with a list of genes (or

DNA segments or regions of the genome) and a list of characters in a sample of people. You now draw up a big table, called a correlation matrix, to find the most significant associations. How often is liver disease associated with the gene *Visigoth*? How often is being good at football associated with *BentSquirrel5*? (I'm making up the gene names ... I hope.) Having done so, you pick the strongest association you can find, and run the relevant data through a statistics package to find out how significant it is. You then declare this association to be statistically significant at the level you have calculated, and publish that particular result, while ignoring all the other pairs of variables that you looked at.

What's wrong? Why does the next study fail to find any such association? Why would we *expect* no confirmation? Because you *chose* a pair of data sets that were unusually closely correlated. You then, in effect, pretended you'd bumped into it at random. It's like sorting through the pack to find the ace of spades, slapping it on the table, and claiming to have achieved a feat with probability 1/52.

Suppose you are looking at 10 genes and 10 characters. That gives 100 pairs. Of those 100 cross-correlations, given random variation, you expect one – on average – to be 'significant at the 99% level' – *even if there is no causal connection whatsoever*. (Actually, those 100 events won't be completely independent. A similar criticism holds if that is taken into account, but the mathematics is less transparent.) If you now use the significance criterion to reject the other 99 pairs, and keep the significance level that the package gives you, there you have the fallacy. Not surprisingly, the next independent trial finds no significant association at all. It was never there.

The correct methodology should be to use one group of subjects to home in on a possible connection, but then to check it using a second, independent group (ignoring all data from the first trial and looking only at associations you've already chosen via the first trial). Often, however, the first study published in a journal and announced to the media carries out only the first step. Eventually a different team carries out the second step ... and, surprise surprise, the result can't be replicated. Unfortunately, it may take quite a while for the second step to be performed, and the mistaken claim to be corrected, because there is little scientific kudos to be gained by repeating other people's experiments.

. . . .

The three billion DNA bases in 'the' human genome may seem a lot, but in terms of computer storage, it constitutes a mere 825 megabytes of raw data. This is much the same as one music CD. So we are roughly as complex as *Sergeant Pepper's Lonely Hearts Club Band*.

Because the information content of the human genome is so small, it is now possible to sequence the genome of an individual at a cost of between $5,000 and $15,000, predicted to drop to $1,000 within a couple of years. (The precise cost depends on how much of the genome is involved and other factors, including why the sequencing is being done and who is doing it.) This 'personal genomics' leads into an aspect of the Human Genome Project that was somewhat neglected in the heady rush to obtain the sequence. There is no such thing as *the* human genome. Different individuals have different alleles (gene variants) at particular genetic locations (such as, but not limited to, hair colour, eye colour and blood type) and also differ in parts of the genome that do not code for proteins, such as the so-called variable tandem repeats, where the same DNA sequence is repeated over and over again. In fact, this is the basis of genetic fingerprinting, which was introduced by forensic scientists as a way to associate DNA traces with their owners. It wouldn't work if all humans had the same genome.

However, we all have much the same basic framework for our DNA, and this is what the Human Genome Project was actually about. As it turned out, Celera's genome really was personal: it was partly based on its founder Craig Venter's own DNA.

In the heady days when the Human Genome Project was first seeking funding, the idea was sold to governments and private investors not as a vital piece of basic science, but as something that would inevitably lead to massive advances in our ability to cure diseases. Once you know 'the information' that makes a human being, surely you know everything about that being. Well no, because you're confusing two different meanings of 'information' – what is encoded in DNA, and what you would need to know to put a human being together from scratch. Similarly, the telephone directory gives you 'the information' you need to get in touch with someone, but you also need a telephone, and you're out of luck if they're away on holiday. (Less so now that we have mobile phones, but you get the point.)

To date, the pay-off from the Human Genome Project, in terms of curing diseases, has been virtually non-existent. This is really no great surprise. For instance, the genetic basis of cystic fibrosis is the gene *CFTR* . It contains about 250,000 base pairs, but the protein that it encodes (cystic fibrosis transmembrane conductance regulator) is a chain of 1,480 amino acids. Get this chain wrong, and the protein doesn't work. In 70% of the mutations seen in people with cystic fibrosis, three specific base pairs go missing. This triplet codes for phenylalanine, at position 508 of the protein. Its omission causes cystic fibrosis. The remaining 30% of cystic fibrosis patients have, between them, about a thousand different mutations of *CFTR*.

Most of this has been known since 1988, but no cure for cystic fibrosis has yet been discovered. Gene therapy, a technique for changing the DNA in the cells of a living human by infecting them with a virus that carries the required sequence, has run into serious trouble after the deaths of several patients. Some forms of this treatment are currently illegal in various countries; however, the technique has had limited success in the treatment of X-linked severe combined immunodeficiency, popularly called bubble-boy syndrome because sufferers have to be isolated from those around them to avoid severe infections.

There is a growing realisation that, with a few standard exceptions, our genes do not cause – or even predict – the diseases that we will contract throughout our lives. The US Government is now taking urgent steps to regulate the activities of personal genomics companies, to prevent the exploitation of inaccurate public perceptions of genes.

As basic science, the Human Genome Project constitutes a huge breakthrough. As a major advance in medicine, it has yet to perform. Even as basic science, its main outcome has been to force major revisions of biologists' previous assumptions about human genetics. I've already mentioned that before the human genome was sequenced, it was believed that there must be about 100,000 genes, in the sense of 'sequences that code for proteins'. The reason was straightforward: the human body has about 100,000 different proteins. As already remarked, it turned out that only 25,000 or so such genes exist. What we then learned is that genes break up into isolated segments, which can be combined in many ways, so the same gene can code for several proteins. The idea that an individual's DNA sequence is some kind of dictionary of its proteins turns out to be naive and simplistic.

All this makes the Human Genome Project excellent science: it changes our views. Unfortunately, the resulting picture has turned out to be more complicated than biologists had expected, and it is becoming clear that the gap between sequencing an organism's DNA and knowing how that organism works is far greater than most people had hoped.

9 Taxonomist, Taxonomist, Spare that Tree

Anyone who visits a zoo quickly notices that some animals are more alike than others. Lions, tigers, leopards and cheetahs are all variations on a basic 'cat' design; polar bears, brown bears and grizzly bears are all bears; wolves, foxes and jackals are dog-like, and so on. Our best explanation of these resemblances, along with misleading exceptions such as dolphins resembling sharks, is evolution. However, the similarities were developed into a systematic classification of life on Earth, long before Darwin devised the first credible theory for their occurrence. One of the first steps in the development of any branch of science is to find a way to organise the wealth of observations that nature presents to us, and this is especially necessary in biology, because of the vast diversity of life.

As we've seen, the first important step in this direction was made by Linnaeus, with his ambitious scheme to classify not just animals, but plants and minerals as well. He was not the first to try to bring some kind of order into such matters, and some of his terminology goes back to Aristotle, but his method was the first to be widely adopted.

We can represent the eightfold hierarchy of taxonomic ranks diagrammatically. Life subdivides into three domains; each domain subdivides into a number of separate kingdoms, and so on. Mathematically, a series of subdivisions of this type has the structure of a *tree* – a diagram with repeated branchings (see Figure 18). The trunk of the tree is Life; this splits into three major

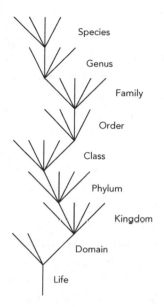

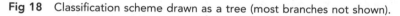

Fig 18 Classification scheme drawn as a tree (most branches not shown).

limbs, the domains, which are eukaryotes, prokaryotes and archaea. Each domain splits into kingdoms: for example the eukaryote domain divides into animalia, plantae, fungi, amoebozoa, chromalveolata, rhizaria and excavata. The first two are animals and plants, the third is what it appears to be, and you can look the fancy ones up if you want to know what they are. Then animalia divides into a large number of phyla – so large that it is normal to first split it into subkingdoms, then these split into superphyla, and finally those split into phyla.

One of the reasons why these further subdivisions have arisen is that, over time, we have discovered vastly more species than were known when Linnaeus first began to catalogue nature's diversity. But the growing complexity of the system of nomenclature, and the arguments that often accompany this process, also indicate that the rich panoply of life does not easily fit into any preassigned scientific straitjacket. Many modern biologists think that this system is no longer adequate to describe the complicated interrelationships found in living creatures, which is probably correct, but it does suffice to *label* them, and it is convenient, traditional and comprehensible by humans – unlike the suggested replacements.

• • • • • • • • • • • •

Linnaeus's classification scheme made it possible to establish, fairly definitively, whether any new-found organism was a genuinely new species or one that was already known. It also had a seductive allure, structuring all Earth's life forms into a single Tree of Life, one of the enduring and iconic evolutionary images. This captures, in diagrammatic form, the relationships among present-day species and their evolutionary ancestors. Ernst Haeckel produced many wonderful evolutionary trees, in a somewhat baroque style of representation; precursors go back to Darwin's notebooks and a diagram in the *Origin*. Darwin, realising that he lacked sufficient data, did not commit himself to a single origin. But it is clear that he did not expect dozens of independent origins for various types of creature. A majestic tree, with bifurcating boughs, branches, twigs and twiglets, provides a vivid metaphor for the idea that all living creatures are related, and that there was a single origin of life (see Figure 19). Or possibly a small number of separate origins, which would lead to several disconnected trees.

Part of the image's appeal is our familiarity with 'family trees', especially of royal families. Today there are genealogical websites where you can research your family's history and draw up a tree showing your parents, grandparents, siblings and other close relatives. This familiarity makes us think we understand family trees, and leads us to view the Tree of Life as something similar. However, diagrams like this can cause confusion. Do the branches represent species, or relationships among species? Do they represent species that exist today, or ones that used to exist and are now found only as fossils? When evolutionary explanations came into fashion, the distinction became vital, but was often ignored. For example, the jibe 'Which ape was your grandfather, Mr Darwin?' assumed that today's apes could be humanity's past ancestors. This isn't what Darwin was suggesting, and in any case it's impossible unless someone invents a time machine.

Is a tree an appropriate metaphor for evolutionary divergence?

When species split through evolution, a process known as speciation, a single species typically becomes two. It is difficult not to speak of this process as 'branching', and just like the branches of a real tree, species split repeatedly. Trees have always loomed large in humans' daily life, so the metaphor is a natural one.

However, it can be pushed too far. Haeckel's tree diagrams

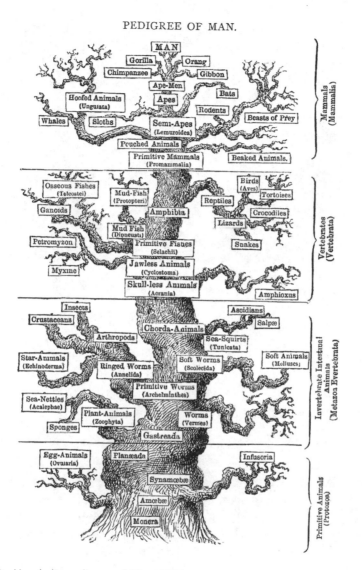

PEDIGREE OF MAN.

Fig 19 Haeckel's pedigree of Man, 1906.

resemble real, though artistically stylised, trees. His artist even gave them bark and roots. And he made the trunk thicker than the branches that split off from it, which is distinctly misleading if, as seems natural, the thickness of a branch represents the abundance of the corresponding species. The real Tree of Life starts with a thin trunk, and many branches are thicker than the trunk from which they grow, as some species become wildly successful and populate the planet in huge numbers. Haeckel also drew his trees so that

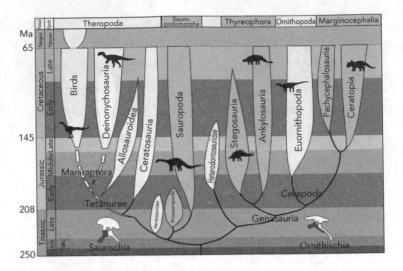

Fig 20 A less misleading Tree of Life showing how birds evolved from dinosaurs.

'higher up' meant 'more advanced'. Superficially, this also correlates with 'more recent', until you realise that branches corresponding to, say, bacteria have to reach all the way to the very top of the tree, and in Haeckel's pictures they don't. For such purposes, a better representation looks less like a real tree, though it still has the characteristic branching behaviour (see Figure 20).

• • • • • • • • • • • •

Mathematicians have their own concept of 'tree', and it too is a metaphor, one that has been enshrined as a specific concept: a diagram in which dots, which can be omitted when the junctions are obvious, are connected by lines, and branches cannot reconnect. This is equivalent to the requirement that no set of edges forms a closed loop. Mathematical trees appear in a more modern realisation of the 'tree' metaphor, known as a cladogram, which comprises little more than the branch points and their timing.

Figure 21 shows a cladogram for the domestic dog and various evolutionary relatives. Note that time runs from left to right here, whereas it runs from bottom to top in Figures 19 and 20. Both conventions are common. The black bear is included deliberately as an 'outgroup', expected on many grounds to be far less closely related to dogs than any other species in the diagram. This is a technical device to permit reliable comparisons, and also a quick-

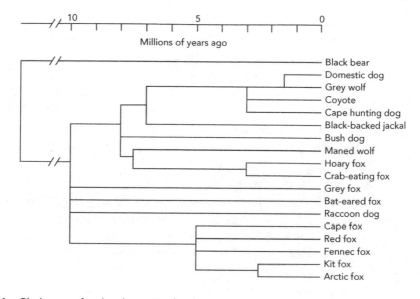

Fig 21 Cladogram for the domestic dog.

and-dirty test of the end result: if dogs turn out to be more closely related to bears than jackals, we would suspect something had gone wrong and re-examine our data. We expect the outgroup to determine the base – the trunk – of the tree.

Cladograms are assembled by computer analysis of similarities and differences between species. These might be lists of characters, such as 'four-legged' or 'has canine teeth'. The timings here are imprecise and give little more than the ordering of successive splits. But it is becoming increasingly common to use lists of alleles (gene variants) or even DNA sequences related to genes, with the timing inferred from the 'genetic clock', the average rate at which mutations occur.

I could say a lot about all this, but I don't want to get sidetracked into a technical area. Suffice it to say that nothing here is guaranteed to be 100% accurate. In particular, the resulting tree is the one that is most likely to fit the data, according to various more or less arcane measures of likelihood. This does not mean that it is definitely an accurate reconstruction of the actual pattern of evolutionary descent. Of course, if more independent data can be collected and a given tree structure survives, that increases the likelihood that it *is* correct. A cladogram is a diagram encoding a long list of statements like 'according to the following criteria, a crab-eating fox is more like a maned wolf than it is like a coyote'.

· · · · · · · · · · · ·

Cladistics was introduced by the entomologist Willi Hennig in 1966, in his book *Phylogenetic Systematics*. As the title suggests, he wanted to make the classification of organisms more systematic, avoiding the often subjective decisions of traditional taxonomy. For Hennig, the basic unit of classification was the *clade*, which consists of an ancestral organism along with all its evolutionary descendants. In a tree diagram, a clade is a single branch together with everything that grows from it.

Conventional cladistics (the construction of cladograms) starts with the assumption that what we are seeking is a tree. If the real pattern of descent is not tree-like, the method will find a tree anyway. This is not as bad as it might seem, because a tree is usually the sensible option. With simple modifications to the method, the tree structure itself can be tested.

The number of possible trees grows rapidly with the number of species. There are, for example, 105 distinct trees for 5 species, and 34,459,425 for 10 species. There is even a formula: for n species, the number of trees is $1 \times 3 \times 5 \times 7 \times \cdots \times (2n-3)$. This is super-exponential growth – faster than any power of a specific number. Somehow, the 'best' tree has to be chosen from among all these possibilities. Naturally, there are many different definitions of 'best', and for any such definition, there are many different mathematical schemes for finding it.

The methods used have become very technical. They are carried out using computers, because the amount of data, and the complexity of the calculations, are greater than an unaided human can handle. But in the early days of cladistics, a lot was done by hand. In simple terms, the technique involves three steps: collect data on the organisms concerned, think about suitable cladograms, choose the best of these. The data take the form of lists of specified characters, so that for bird species it might be things like width of beak, length of beak, colour of feathers, size of feet. Once DNA sequencing became practical (initially for short sequences such as mitochondrial DNA) the data collected usually include DNA, and today many practitioners use nothing else.

The mathematical task is now to find which tree fits the data best. This requires defining some number, known as a metric, that quantifies how closely the tree agrees with the data. Two species with similar data should be close together in the tree – that is, their common ancestor should not be many branches back. Species with less similar data should be separated by more branches. The actual recipes are not as vague as this outline might suggest. Also, there are well-established guidelines for avoiding a choice of characters that might be misleading – something that arises in many different organisms for reasons that do not relate to common ancestry. The shapes of sharks and dolphins, with the same sort of tail and triangular dorsal fin, are examples.

Suppose for example, that we are trying to sort out the relationships among four species of cat: (domestic) cat, leopard, tiger and cheetah. To keep ourselves honest we include an outgroup: snail. We select four characters (way too small for a serious analysis, but it will show how the method works), and tabulate these against the five species, using 1 for 'yes' and 0 for 'no', as in Table 6.

As a measure of how closely different species are related (which is the opposite of 'distance', so minimising distance is the same as maximising closeness), we could use the number of entries in the matrix that they have in common. For example, cat and leopard agree on whiskers and purr, but not on spots and big, so the distance is 2. In this case the small quantity of data lets us tabulate all possible closenesses, as in Table 7.

Next, we apply some heuristics, a fancy word for 'informed guesswork'. The closest any two species get is 3, and that is for all four types of feline, so we guess that the four types of cat are more closely related to each other than to the snail. This places the snail at the bottom of the tree, where it belongs. Next, the cat is closely related to the tiger and cheetah (closeness 3), but less closely to the leopard (closeness 2), so we expect to find these three at the top of the tree. So we make the cheetah the first species to branch away from snail. Among cat, leopard and tiger, the first two are closer to the cheetah, so we make the tiger branch before the other two do.

Table 6 Four characters in five species.

	Whiskers	Spots	Purr	Big
Cat	1	0	1	0
Leopard	1	1	1	1
Tiger	1	0	1	1
Cheetah	1	1	1	0
Snail	0	0	0	0

Table 7 Closeness of the five species.

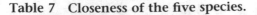

Cat/leopard	2
Cat/tiger	3
Cat/cheetah	3
Cat/snail	2
Leopard/tiger	3
Leopard/cheetah	3
Leopard/snail	0
Tiger/cheetah	2
Tiger/snail	1
Cheetah/snail	1

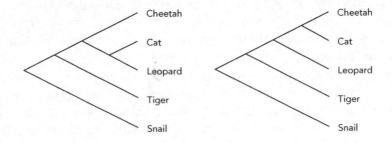

Fig 22 Two candidate cladograms.

At this point there are two different ways to complete the tree, illustrated in Figure 22: either cat and leopard both branch from the line leading to cheetah, and then split, or cheetah and cat branch from leopard.

In the first picture, cat is closer to leopard than to cheetah, but actually it's closer to cheetah than to leopard, which is what the second picture shows. So we plump for the second tree. But this tree is not the answer: just the first step in finding it. If we had made comparisons in a different order, for instance, we might have been led to a different tree. To complete the construction of the best-fitting tree, we therefore need some measure of how well the overall tree fits the data. Then we look at variations on our candidate tree – say, swapping leopard and tiger – and see whether they do better. We should also swap snail with, say, tiger, to make sure that our outgroup really is an outgroup, otherwise there's no point in including it in the first place.

To find the best tree, we need to work out the sums for all possible trees. The formula tells us that there are 15 of these, so the calculation is feasible, but in practice we work with many more characters, and a different approach is needed, described below. Since this example is much too simple-minded anyway, I won't take the analysis any further, but the general gist of the method should now be apparent.

Because the number of possible trees grows very rapidly with the number of species, it is not possible in practice to calculate the best-fitting tree with perfect accuracy. However, many methods exist to find a tree that fits almost as well as the theoretical best one. They are borrowed from an area of mathematics known as optimisation, often used in industry and economics.

The construction of a plausible cladogram might go like this. First, the cladist uses their experience, or other so-called heuristic methods, to write down a small number of trees that are expected to be close to optimal. These are input into a computer running suitable software, and the computer randomly generates trees that are slight modifications of the initial guesses. It then calculates the metric – how good the fit is – and sees which of the modified trees performs best. The process is now repeated, with random variants of this new tree, and continues until no random modification makes the tree any better.

In an analogy, imagine that the metric represents height in a landscape. The best-fitting tree corresponds to the highest point in the entire landscape. However, there may be several hills, each with its own local peak; only one of those will be the highest point around. So the idea is to choose a few plausible starting points, and then search randomly near those to see if any path leads upwards. If so, wander a little way up the path and repeat the search. The main problem with such methods is that if your initial guess isn't a good one, you may get stuck on a hill that is not the highest one around. Searching nearby won't improve the outcome; you have to search further afield. There are sensible ways to do that, but none of them are foolproof.

There is also no guarantee that the tree obtained by this method is actually the precise evolutionary tree of the species involved. But we can be fairly confident that if the tree shows two species to be very closely related, or very distantly related, then much the same holds for the genuine evolutionary tree. We can be very confident if different data, analysed by different methods, lead to similar results.

• • • • • • • • • • • •

All very well, but ... How sensible is it to model evolution as a tree?

In *Wonderful Life*, Stephen Jay Gould discussed the diverse soft-bodied creatures found in fossils from the Burgess Shale, a deposit of rock strata in Canada. These deposits, and the fossils within them, come from the time of the Cambrian explosion, a sudden burst of diversity that led to the evolution of many different, highly complex creatures. Unusually, the fossils preserve many soft features that would normally have rotted away. According to Gould's interpretation, the evolutionary descent of the Burgess Shale fossils looks more like a bushy savannah than a single tree. However, bushes corresponding to species that have all become extinct do not reach to the present day, and so cannot be reconstructed from present-day data.

In fact, Gould suggested that the Burgess Shale fauna contained more phyla – one of the largest units into which life forms are classified – than currently exist. Humans, for example, belong to the phylum of chordates, creatures that develop a notochord as an embryo. He went on to deduce that the evolution of humanity involved a random 'accident' at the time of the Cambrian explosion. *Pikaia*, which among the Burgess Shale fauna is the best candidate ancestral species for all chordates, left surviving descendants. *Anomalocaris*, *Opabinia*, *Nectocaris*, *Amiskwia* and various other organisms, each representing a distinct (and now extinct) phylum, did not – even though all these creatures were happily coexisting, and there seems to be no good reason to expect any one of them to survive and the others to die out.

It now seems that Gould inadvertently exaggerated the differences among the fossils he considered, and many are in fact related to existing creatures, contrary to what he thought. However, it is also true that many equally baffling Burgess Shale fossils have not yet been analysed at all, so Gould's theory might yet be revived. At any rate, if you look for a tree you will find a tree, so for some questions it makes sense not to start by assuming that a tree exists.

• • • • • • • • • • •

In a genetic interpretation, the Tree of Life represents how genes pass from (organisms in) ancestral species to (organisms in) their descendant species. However, there is a second way for genes to be transferred between organisms. It was discovered in 1959 by a Japanese team, which discovered that antibiotic resistance could be transmitted from one species of bacterium to a different species.[1] This phenomenon is known as horizontal (or lateral) gene transfer, whereas the conventional transmission of genes to descendants is called vertical transfer. The terms are derived from the usual tree diagram of evolution, with time running vertically and species-type horizontally, and have no other significance.

It soon became apparent that horizontal gene transfer is widespread among bacteria, and not uncommon in single-celled eukaryotes. This changes the paradigm for evolution among such creatures, because it introduces a different way for genomes to change. The classical concept of genetic changes arising through mutations (including deletions, duplications and reversals, as well as point mutations) in the genome of an organism in a single species must be broadened to allow the insertion of segments of DNA from a different species altogether. There are three main mechanisms for such transfer: the cell may incorporate alien genetic material through its own workings, alien DNA may be brought in by a virus, or two bacteria may exchange genetic material ('bacterial sex').

There is also some evidence that multi-celled eukaryotes may have been the recipients of horizontal gene transfer at some stage in their evolutionary history. The genomes of some fungi, in particular yeast, contain DNA sequences derived from bacteria. The same goes for a particular species of beetle that has acquired genetic material from *Wolbachia* bacteria, which live inside the beetle in a state of symbiosis. Aphids contain genes from fungi which let them manufacture carotenoids. The human genome includes sequences derived from viruses.

These effects certainly change our view of how genetic changes, one of the driving forces behind evolution, can occur. They imply that many creatures' genetic ancestry involves more than their obvious evolutionary ancestors. A number of biologists have argued that this forces us to abandon the Tree of Life metaphor. Scientifically, this poses no great obstacles: the Tree of Life is not sacred, and if the evidence indicates that it is wrong, it should be discarded. Our view of evolution would then be different – at least in so far as the standard metaphor goes – but science often progresses by revising previous ideas. So does horizontal gene transfer wreck the Tree of Life metaphor?

At first sight the answer seems to be 'yes'. Horizontal gene transfer can introduce closed loops, by linking two distinct branches of the usual tree. And then it's not a tree.

However, the branches in Haeckel's Tree of Life, and in cladograms, represent how *species* branch, either historically or conceptually. They don't represent individual organisms. Horizontal gene transfer moves a snippet of DNA from one organism to another. So this new link is not a branch on the species tree. A cow becomes a cow with a bit of alien DNA, but it's still a cow. Of course, the alien DNA affects what it might evolve into in future, but that comes later, if at all.

In a diagram with conceptual branches showing how organisms or species are connected by changes to their DNA, horizontal gene transfer does throw in some extra connections that spoil the tree structure. But this doesn't mean that the original Tree of Life metaphor was wrong. It just means that we're talking about a different metaphor.

In short, horizontal gene transfer has no effect on the Tree of Life for *species*. It has a small effect on the tree for organisms, and a bigger effect on the tree for DNA. There is perhaps one exception to these statements: when the species are bacterial or viral. Then horizontal gene transfer is so common that even the concept of a species is questionable.

Speciation, considered for individuals, is probably a very complicated intermingling of edges. Representing the speciation event as a simple branch point almost certainly oversimplifies the process, and leads to questions and distinctions that may not be appropriate (such as 'exactly when did the two species split'?). Some complex system models of speciation introduced by Toby Elmhirst under the name BirdSym exhibit very complex cascades of changes in phenotype during speciation events. The pictures look more like braided rivers than simple branch points.

. . . .

Might there be not one Tree of Life, but several? Darwin left this possibility open in the *Origin*. The general idea of evolution would not be greatly affected, whatever the answer, but there is an evolutionary reason to prefer a single tree. Once life gets going – by whatever method – it reproduces, and this makes any subsequent independent origin unlikely to get very far. The new kids on the block have to compete with those that are there already, who have an advantage because they have become pretty good at playing the evolutionary game. So we would expect a single origin, and would need new ideas to explain a multiple one.

In 2010 Douglas Theobald used methods from cladistics to test this hypothesis, known as 'universal common ancestry', and the results came down firmly in favour of a common ancestry for all present-day life.[2] The word used here is 'ancestry', not 'ancestor', with good reason. Theobald's model permits the last universal common ancestor to be a population of different organisms, with different genetics, living at different times. His method involves the amino acid sequences of 23 proteins, found across all three domains of life – archaea, prokaryotes and eukaryotes. You can think of them as molecular probes that cover the entire range of living creatures, and go way back into deep time. Having chosen the proteins, the next step is to calculate evolutionary trees and sets of several trees. The final step is to compare how likely these results are, given the data.

Theobald compared a single tree, perhaps with additional horizontal gene transfer, with (say) two trees, which may or may not be linked together by horizontal gene transfer. His result is dramatic: a single tree is about $10^{2,860}$ times as likely as two or more trees (see Figure 23). To put this in perspective, it is like randomly shuffling a pack of cards and finding that the cards are arranged in perfect order, ace to king for spades, hearts, clubs and diamonds ... and repeating this 42 times.

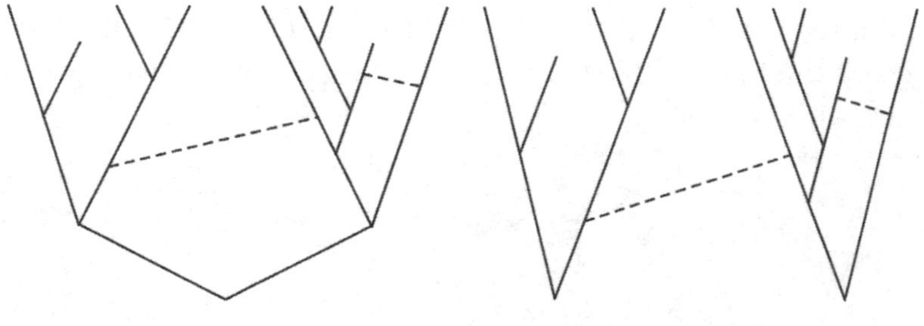

Fig 23 A unique Tree of Life like the left-hand diagram is $10^{2,860}$ times as likely as a multiple tree like the right-hand one. Dashed lines represent horizontal gene transfer and are not part of the tree.

10 Virus from the Fourth Dimension

• •

Geometry became a well-developed and powerful branch of mathematics through the work of ancient Greek philosophers and mathematicians. The most famous of the ancient Greek geometers, though not the most talented, was Euclid of Alexandria.[1] He put geometry on a systematic basis in his *Elements*, a logical development of geometry written around 300 BC. It became the most successful textbook ever written, with thousands of editions since it first saw print in Venice in 1482.

The climax of the *Elements* is the classification and construction of the five regular solids: the tetrahedron, cube, octahedron, dodecahedron and icosahedron. The names, cube aside, refer to the number of faces: 4, 6, 8, 12 and 20, respectively.[2] The cube has square faces, the dodecahedron has pentagonal ones and the other three are made from equilateral triangles (see Figure 24).

In most areas of science, discoveries made 2,300 years ago are no longer terribly relevant – although Archimedes' principle about floating bodies and his law of the lever still have their uses, and

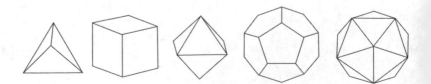

Fig 24 The regular solids. *Left to right*: Tetrahedron, cube, octahedron, dodecahedron, icosahedron.

they are about that old.[3] But mathematics builds new discoveries on top of previous ones, and once something has been proved true it remains that way. So it tends to hang around. New standards of rigour come into being, definitions are made more watertight, new interpretations are introduced, and topics that were once flavour of the month may sink into obscurity, but fundamental mathematical ideas are pretty much permanent.

The icosahedron is a case in point. It has always played a role in pure mathematics: in 1908 the French mathematician Charles Hermite discovered how to use the geometry of the icosahedron to solve algebraic equations of the fifth degree. But until the twentieth century there were no significant applications of the icosahedron to the real world, because it did not seem to arise in nature. Since 1923 it has appeared in the design of objects: the engineer Walther Bauersfeld used it as the basis of the first planetarium projector, and the architect Buckminster Fuller reinvented the idea and used it to design geodesic domes. It also underlies the geometry of the modern football, which is an elaboration on the icosahedron: you truncate it – cut off the corners – to make it more rounded (of course). The same elaboration captures the structure of buckminsterfullerene, a recently discovered molecule consisting of 60 atoms of carbon and nothing else, which form a roughly spherical cage (see Figure 25).

Fig 25   *Left*: Icosahedron. *Middle*: Truncated icosahedron and football. *Right*: Buckminsterfullerene.

When electron microscopy and X-ray diffraction methods got going, however, Euclid's 20-faced solid became a regular feature of biology. The context was viruses – diminutive structures too small to be seen in an optical microscope, but visible in a more powerful electron microscope. Viruses are a major cause of diseases in humans, animals and plants; the Latin word 'virus' means 'poison'. Viruses are a bit larger than most biological molecules, but a typical virus is about one-hundredth the size of a typical bacterium. Since volume varies as the cube of length, you could pack a million ($100 \times 100 \times 100$) viruses inside a single bacterium if no space were wasted. There are about 5×10^{30} bacteria on Earth, but the number of viruses is about ten times that. Neither of these figures is particularly accurate, and both may be underestimates, but they give a general feel for the numbers. Viruses outnumber humans 10^{22} to one.

Bacteria definitely qualify as life, because they can reproduce using their own genetic processes. Viruses may or may not qualify: they have genes – DNA or RNA sequences – but they cannot reproduce using just their own genes. Instead, they reproduce explosively by subverting the reproductive biochemical machinery of a bacterium, much as a document can reproduce through the intermediary of a photocopier. (Actually, a few viruses can replicate unaided, but these are exceptional.) Some biologists argue that the definition of life should be extended to include viruses.

Over 5,000 distinct types of virus have been found since Martinus Beijerinck made the first discovery, of the tobacco mosaic virus, in 1898, and indirect evidence tells us that there must be millions more. Most viruses have two main components: genes formed from DNA or RNA, enclosed within a protein coat known as a capsid. The capsid is typically formed from identical protein units, known as capsomers. Some viruses also possess an extra layer of lipids (fat) for protection when outside a cell.

As early as 1956, it was noticed that the majority of viruses are either icosahedral or helical: shaped like a football or shaped like a spiral staircase. Some have a more complex structure: for example, enterobacteriophage T4 has an icosahedral head, a helical stalk and a hexagonal baseplate from which fibres extend. It looks like a lunar landing molecule (see Figure 26). But the main form observed is Euclid's elegant icosahedron, which, devoid of practical application for more than two thousand years, turns out to be just right for making a virus (see Figure 27).

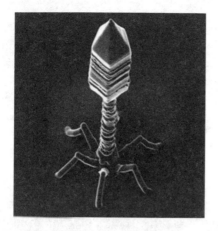

Fig 26   Enterobacteriophage T4.

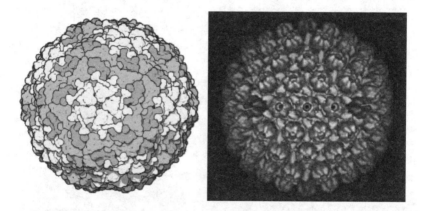

Fig 27   Icosahedral structure of two viruses. *Left*: Foot-and-mouth disease virus differs from its mirror image and has 60 symmetries. *Right*: Herpes simplex virus is mirror-symmetric and has 120 symmetries.

There is a hand-waving explanation based on minimising energy, which goes like this. Virus coats are typically constructed from many copies of a roughly spherical protein molecule. A collection of such molecules has the least energy – something that nature finds desirable – if it is as close as possible to being a sphere. Soap bubbles *are* spheres, because this is the shape that has the least surface area, hence the least energy of surface tension, while enclosing a given volume. Virus coats can't form exact spheres, because the component protein units cause bumps. (Try to fit a hundred tennis balls together to make a smooth sphere.) So they do the best they can. Of all Euclid's solids, the icosahedron is closest to a sphere. The truncated icosahedron is even more spherical, hence its use for footballs (and as a bonus, the panels become even rounder when the ball is inflated). So evolution and FIFA (the Fédération Internationale de Football Association) independently came up with the same shape, for the same reason. Until the 2010 World Cup, that is, when the balls were made in a different way – and everyone complained.

. . . .

At the heart of the icosahedron, indeed of all five regular solids, is the concept of symmetry. Since the early 1800s, mathematicians have developed a profound theory of symmetry, with applications throughout the sciences, known as group theory. I could spend a book discussing how group theory came about (and I have done[4]). The key point is that a symmetry of an object is not a thing, but a *transformation*, whose application leaves the object looking exactly the same.

Put baldly, that sounds like gobbledegook: a symmetry transforms the object but leaves it looking exactly the same. Yes, and there are little green men on the Moon but they're invisible so nobody can ever detect them ... Right. Actually, the statement makes sense if properly construed. A transformation is a way to rearrange things or move them around. In this case, the relevant transformations are rigid motions, especially rotations and reflections. Now, think of a square, which intuitively has quite a lot of symmetry. For example, all four corners are the same shape. One way to capture this feature is to rotate the square about its centre through a right angle. The result is an identical square, in an identical orientation. If you shut your eyes while the square was being rotated, and it had no markings on it, then when you opened them again you wouldn't notice that anything had happened. So the transformation 'rotate through a right angle' is a symmetry of the square. In all, there are eight such symmetries: leave the square alone; rotate it through one, two or three right angles; reflect it in either diagonal; reflect it in either line joining the midpoints of opposite sides.

These transformations have a pleasant kind of 'closure': perform any two of them in turn, and the result is one of the same eight symmetries. They are said to form a group. The same is true for the symmetries of any other object. A circle is much more symmetric than a square, and has infinitely many symmetries: rotate through any angle, reflect in any diameter. But again, any two symmetries performed in turn have the same effect as a single symmetry. Rotate by 14° and then rotate by 53°: the result is the same as a single rotation through 67°, which is 14°+53°.

Euclid's regular solids have rich and beautiful symmetry groups. The tetrahedron has 24 symmetries, the cube and octahedron have 48, and the dodecahedron and icosahedron have a massive 120. The symmetry properties of these solids are what make them so prevalent in modern pure mathematics. Notice that different solids here, such as the dodecahedron and icosahedron, can have the same number of symmetries. There is a good reason: if you draw a dot in the middle of each face of an icosahedron, you get the 20 vertices of a dodecahedron. Similarly, if you draw a dot in the middle of each face of a dodecahedron, you get the 12 vertices of an icosahedron. Suitable geometric relations between shapes can give them the same symmetry groups.

* * * * * * * * * * * * *

The architecture of the viral capsid is important biologically: in particular, it helps with the analysis of images of the virus (such as those obtained by X-ray crystallography) and the construction of models of how the virus assembles. The icosahedral structure of viruses does not just determine the overall shape: it is inherent in the arrangement of the protein units. Until recently the main theoretical description of capsid architecture was the one derived in 1962 by the American and British biophysicists Donald Caspar and Aaron Klug.[5]

Icosahedral virus coats are made from triangular arrays of capsomers, fitted together like the faces of an icosahedron. Each triangle is made from row upon row of capsomers, arranged like the balls at the start of a game of snooker or pool. Looking more closely still, that's not quite the full story: the rows of capsomers can be skewed, so that some rows run into the edge of the triangle and warp across it into the next triangle. Nature obviously has no trouble making such shapes, but they are slightly odd mathematically, and the first step in understanding the patterns is to work out their mathematical status and find their common features.

Like pool balls, most capsomers are surrounded by six others (hexamers). However, some are surrounded by only five others (pentamers). It turns out that this condition is forced by geometry. If we represent the virus capsid as a polyhedron by placing vertices at the capsomer and joining adjacent capsomers by edges, then hexamers are vertices lying on 6 edges, and pentamers lie on 5 edges. This fact alone imposes mathematical conditions on the possible number of capsomers. Leonhard Euler, one of the all-time mathematical greats, discovered a formula relating the number of faces, edges and vertices of a solid. Namely, for any polyhedron topologically equivalent to a sphere,

$$F - E + V = 2$$

where F is the number of faces, E the number of edges and V the number of vertices. For example, a cube has $F=6$, $E=12$, $V=8$, and $6-12+8=2$. This general fact is known as Euler's formula for polyhedra. A simple calculation using the formula shows that in any virus coating composed purely of hexamers and pentamers, there must be exactly 12 pentamers. The method doesn't specify where they occur, but it proves that they have to be present.

Caspar and Klug followed up these topological clues. They looked first at helical viruses, and then went on to consider icosahedral ones. Here the basic mathematical problem is to pack identical units together to form shapes that are close to spherical, bearing in mind that the relation between each unit and those adjacent to it is likely to be restricted by the available chemical bonds. The simplest case is when there is only one such relationship; geometrically, this means that each unit is surrounded by exactly the same configuration of adjacent units. This in turn implies a high degree of symmetry, which for brevity I'll call 'perfect symmetry', and immediately suggests considering the regular solids. Among these, the icosahedron is the most plausible candidate: of all the regular solids, it forms the best approximation to a sphere. In addition, electron microscope images of several viruses appeared to be icosahedral, although Caspar and Klug note that this 'does not necessarily mean that the symmetry down to the molecular level is icosahedral'.

With the requirement of perfect symmetry, an icosahedral

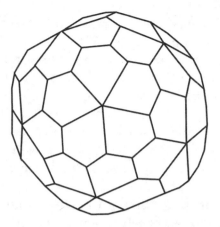

Fig 28 Fitting 60 identical units together so that each has the same relation to its neighbours.

arrangement can clearly accommodate 12 or 20 units: 12 if they are situated at the corners, 20 if they are placed at the centres of the faces. The largest number of units that can fit together so that each unit has the same immediate neighbourhood is 60 (see Figure 28). This increases to 120 if mirror images are considered identical, but biological molecules tend to have a specific 'handedness', so this is unlikely. Again, the symmetry is icosahedral.

It follows that the number of units in a virus with perfect symmetry must be 12, 20 or 60. However, no viruses known to Caspar and Klug employed these numbers, and most of them had more than 60 units. Indeed, none of them had a multiple of 60 units, which might be realised by relaxing the requirement of perfect symmetry a little. The most likely way out is to relax the requirement even further, and Caspar and Klug found inspiration from an unusual source: the architect Buckminster Fuller. Fuller had a liking for geometric forms, and the geodesic dome is one of his more famous ideas, a roughly spherical enclosure made by fitting a large number of triangular panels together. Such a dome featured as a pavilion for the 1964 New York World's Fair, and hemispherical versions can be found at the Eden Project in Cornwall.

You can't make a geodesic dome from equilateral triangles arranged six to a vertex, because they would form a flat plane. Fuller, following various predecessors, realised that triangles that are *nearly* equilateral can fit together to make a dome. Such arrangements do not have perfect symmetry; instead, triangles have two different kinds of neighbourhood. Consistency with Euler's formula demands that some must be arranged so that five of them fit round a vertex, while the rest fit six to a vertex. Caspar and Klug realised that although adjacent units are generally held together by the same arrangement of chemical bonds, these bonds can be bent by a small amount, so that the bond angles can be slightly different for units that are not symmetrically related. Experiments performed by the Nobel Prize-winning chemist Linus Pauling suggest that bond angles can be changed by about 5° from their average values, which allows some flexibility.

Caspar and Klug were led to an unorthodox range of solids called pseudo-icosahedra, familiar to expert geometers but not to most mathematicians. They are solids that resemble icosahedra but are less regular. They can be constructed from a tiling of the plane (see p. 154) by equilateral triangles. First, choose two numbers a and b (see Figure 29). Starting from a vertex, move a units to the right and b units at 120° to this direction to get a second vertex; then locate the third vertex to form a large equilateral triangle containing many vertices of the original tiling. Twenty of these triangles can then be fitted together to form an icosahedral polyhedron with $10(a^2+ab+b^2)+2$ vertices, of which 12 are pentamers and the rest hexamers. The pentamers always lie on the axes of icosahedral symmetry (the 'corners'). Examples of pseudo-icosahedral architecture are given in Table 8.

The Caspar–Klug theory applies to many different icosahedral viruses, but there are exceptions. Forty years ago, Nicholas Wrigley noticed that some icosahedral viruses do not have this pseudo-

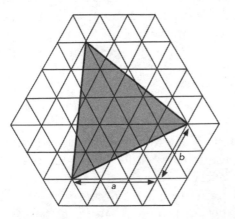

Fig 29 Construction of basic triangle of a pseudo-icosahedron.

Table 8   Numerology of pseudo-icosahedral viruses.

$\{a, b\}$	No. of capsomers	Virus
$\{1, 1\}$	32	Turnip yellow mosaic
$\{2, 0\}$	42	Bacteriophage ΦR
$\{2, 1\}$	72	Rabbit papilloma
$\{1, 2\}$	72	Human wart
$\{3, 0\}$	92	Reo
$\{4, 0\}$	162	Herpes, chickenpox
$\{5, 0\}$	252	Adenovirus type 12
$\{6, 0\}$	362	Infectious canine hepatitis

icosahedral architecture. Instead, they can be described by so-called Goldberg polyhedra, which are hexagonal packings on the surface of an icosahedron.[6] However, even these structures are insufficient to classify the arrangements of capsomers in icosahedral viruses: for example, in 1991 Robert Liddington and colleagues pointed out that polyoma virus has many more pentamers than the 12 found in pseudo-icosahedra and Goldberg polyhedra.[7] So some more general mathematical description was needed.

By now, biologists' minds were on other things, but mathematicians were still puzzling over the exceptions. Around the year 2000 the German-born mathematician Reidun Twarock and her research team at the University of York developed a more general theory of the geometry of viruses based on symmetry principles closely analogous to the group theory of the icosahedron.[8]

There was only one difference: now the geometry took place in four dimensions, not three.

. . . .

The fourth dimension ...

It sounds like something from a science-fiction story, a hidden realm adjacent to our everyday world in which all manner of strange creatures lurk ... Indeed, this is how the concept is portrayed in H.G. Wells's *The Time Machine* of 1895, in which the Time Traveller is taken to the far future of humanity, now separated into the languid Eloi and the grotesque Morlocks. But Wells based his novel on a topic from real science that was becoming popular at the time. He credited the idea to 'student discussions in the laboratories and debating society of the Royal College of Science in the [eighteen-]eighties.' As the story opens, the Time Traveller invokes the fourth dimension to explain why such a machine is possible:

> There are really four dimensions, three which we call the three planes of Space, and a fourth, Time. There is, however, a tendency to draw an unreal distinction between the former three dimensions and the latter, because it happens that our consciousness moves intermittently in one direction along the latter from the beginning to the end of our lives ... But some philosophical people have been asking why *three* dimensions particularly – why not another direction at right angles to the other three? – and have even tried to construct a Four-Dimensional geometry. Professor Simon Newcomb was expounding this to the New York Mathematical Society only a month or so ago.

It is a matter of historical record that Newcomb, one of the most eminent American mathematicians of his day, published on the topic of four-dimensional space from 1877, and he spoke about it to the New York Mathematical Society in 1893. Four dimensions (and more) were important research topics in mathematics and physics.

Some Victorian theologians saw 'the' fourth dimension as a convenient location for God, contiguous with our universe at every point, yet outside it, and affording the Deity a complete view of the entire universe in a single glance. But then the hyperspace theologians decided that the fifth dimension would be even better, the sixth better still ... and that nothing short of the infinitieth dimension would serve for an omnipotent, omnipresent Deity. At much the same time, spiritualists found the fourth dimension equally suitable as a home for the spirits of the dead; believers in ghosts had a similar view, with the obvious change in the identity of the entities that inhabited this hidden dimension; various cults and pseudoscientific bodies found effective ways to rationalise their own beliefs by throwing in a few references to the fourth dimension; and a few outright crooks used topological trickery to 'prove' that they had access to the fourth dimension and con people into giving them money.

It was the mathematicians' fault, really – they had set this particular ball rolling, and physicists picked it up and ran with it. Then popular culture, unrestrained by the need to stay close to reasonable speculation, pushed the idea to the limits – much as the more recent media image of cloning was populating the planet with exact copies of human beings at a time when biologists were failing to duplicate a cell. With Dolly the Sheep, fact began to catch up with fiction, but cloned humans do not yet exist.

The fourth dimension has fared much better, and if anything, fact is now ahead of fiction – except when it comes to time travel. The concept has been entirely respectable in mathematics and physics for over a century, and scientists now routinely employ mathematical concepts of any number of dimensions – four, ten, a hundred, a million. Even infinity. The imagery of multidimensional spaces has spread to biology and economics, together with the associated mathematical techniques. The idea may sound outlandish, but it is actually very natural. Its main relevance to this chapter, though, is a more direct application of four-dimensional geometry to, of all things, viruses. Strange though it may seem, the fourth dimension and its higher cousins are providing important insights into how viruses pack their protein units together.

· · · · · · · · · · · ·

When mathematicians start talking about familiar terms in ways that make no sense, it usually turns out that they have either appropriated the word and given it a totally different meaning, or they have extended the usual meaning to a wider context. A group is not merely a collection of similar objects, a ring has nothing to do with jewellery or even with circles, and you won't find sheep grazing in a mathematical field. Terms like 'dimension', 'space' and 'geometry' fall into the second category, and are easier to misunderstand because the new meaning is not so obviously different from the old one.

The unwritten rule for extending the meaning of a word is that it should retain its original meaning in its original context. It is all right to introduce a new concept of 'space', for example, as long as the familiar spaces of Euclid's geometry – the plane and, well, space – are still included. Provided this convention is obeyed, you won't get confused by applying the new meaning in the old context. Though you might get confused in the new context if you assume that specific features of the old one still hold good – such as 'space' being something that humans can or do live in.

'Dimension' is a case in point. The plane has two dimensions, our familiar notion of space has three. Any extension of the word to other 'spaces' should preserve those facts. However, the traditional definition – the number of independent directions – is not sacred. It's not even sacred in the traditional contexts: you can change the definition as long as the answers are still two and three respectively.

Traditionally, we sometimes speak of individual 'dimensions'. Length is a dimension, so is width, so is height. Some care is needed, because the same word can also mean 'size': the longest dimension (in that sense) of a box is whichever of these three is biggest, but actually the longest 'dimension' is the diagonal, which is bigger than any of them. Mathematics and science have settled on a more general notion of dimension in which we can safely say that some space has, say, ten dimensions, *without* saying what these dimensions are. The emphasis has shifted: the space has dimension (or dimensionality) 10. We don't define things called dimensions and count them to show there are ten of them. That said, lists of ten things do exist – but we don't call the things dimensions. We call them coordinates.

Mathematicians did not introduce spaces of many dimensions for fun, or to impress people. They did so because they needed them. By the end of the nineteenth century, a variety of developments, motivated by everything from pure geometry to celestial mechanics, all seemed to point towards the same new idea. At much the same time, physicists started to realise that many key discoveries made more sense if they were formalised within a 'space-time' of four dimensions: the three traditional dimensions of space, plus an extra one of time. But time was not *the* fourth dimension: just one possibility.

To cut a long story short: the dimension of a space is the number of independent coordinates needed to specify the things that belong to it. Spaces with many dimensions provide a convenient way to describe systems in which many distinct variables can be set to whatever values we desire. The 'space' of all such choices has a natural structure – a direct generalisation of the familiar mathematics of two and three dimensions. In particular, we can specify what it means for two 'points' in such a space to be close together: corresponding variables should have values that are close together.

Moreover, the 'points' need not actually be *points*. The plane is a

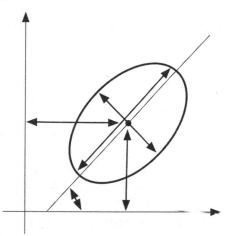

Fig 30 Five numbers are required to determine an ellipse in the plane.

set of points, but it can also be thought of as, say, a collection of ellipses. The set of all ellipses in the plane is an interesting mathematical object in its own right. How can you specify an ellipse? Let's do it in Euclidean geometry, where the pictures are more familiar. You need to know:

- where the centre of the ellipse is (2 numbers),
- how long it is (1 number),
- how wide it is (1 number), and
- at what angle it is tilted (1 number).

So in all, it takes *five* numbers to specify an ellipse (see Figure 30).

The 'space' of ellipses is five-dimensional. And it *is* a space, in the sense that if you change the numbers representing some ellipse by small amounts, you get an ellipse that is 'nearby'. It looks very similar. And the smaller the change, the more similar it looks.

From one point of view, the plane is two-dimensional. From another, it is five-dimensional. But it's the same plane either way, so it makes no sense to maintain that a two-dimensional space exists but a five-dimensional one doesn't. Here, they are two aspects of the same thing. Aside from familiarity and tradition, there is no good mathematical reason to prefer the set of points to the set of ellipses. Which viewpoint is best depends on what questions you're asking. It was for this kind of reason that mathematicians came not only to tolerate spaces of higher dimensions, but to feel lost without them.

• • • • • • • • • • • •

This simple idea turned out to be so useful that it rapidly invaded physics. Today's particle physics, for instance, cannot even be set up properly without using spaces with pretty much any number of dimensions. Engineers got in on the act, too. If you are trying to calculate the stresses and strains in a grid of 100 metal girders, then you have 100 forces to work with. Since you don't know what they are until you do the sums, you are conceptually looking at lists of 100 arbitrary numbers and trying to select the correct one. That is, you are seeking a point in a space of 100 dimensions.

Engineers find that terminology off-putting, and prefer the more physical concept of 'degrees of freedom'. How many different things can vary independently? But it's the same idea. Thinking of all possible configurations of some complicated system as a 'point' in a 'space' of all potentially possible configurations gives such a vivid image, and such a conceptual boost, that it has pervaded every branch of science – and beyond.

A case in point is DNA-space. A sequence of (for simplicity) ten DNA bases permits four choices (A, C, G, T) at each separate location. So 'DNA-space', the ensemble (as physicists say) or set (as mathematicians say) of *all possible* such sequences can be thought of as a 'space' of ten dimensions, with each individual dimension taking four possible values: A, C, G or T. Replace 'ten' by any other number, such as a million, and the same holds good.

This space has a natural geometry. Two sequences are close to each other if they differ in a small number of locations. For instance, AAAAAAAAAA is very close to AAAAACAAAA, a bit further from AAAAACTAAA, further still from AAGAACTAAA, and so on. The 'distance' between two sequences is the number of bases at which they differ. This notion resembles distance in the two-dimensional plane or ordinary three-dimensional space in many respects, although it differs in others. If you are interested in the genetic basis of evolution, whose simplest manifestation is 'point mutations' at which one base changes, this notion of distance is ideal: it equals the minimum number of mutations that could lead from one sequence to another.

Biologists have found the concept of DNA-space, or sequence space, very informative. It matches a similar idea used in computer science to describe digital messages in information theory. Biologists are not alone. Economists view the prices of a million goods as a point (or 'vector') in a hypothetical space of a million dimensions, and the mathematical processes they use, such as optimisation methods, are explicitly motivated by this image. Astronomers, discovering they need 18 numbers to describe the state of the Earth–Moon–Sun system, work in an 18-dimensional mathematical space. The geometry of this 18-dimensional space tells them a lot about how such a three-body system behaves.

The propensity of viruses to undergo genetic mutations at the drop of a hat is informed by this kind of use of multidimensional spaces. It formalises the use of similarities in DNA sequences to infer past evolutionary changes, and it classifies viruses into 'strains', different variants that arise by mutation or exchange of genetic material. These changes are important in medicine, because vaccines that work for one strain may be ineffective for a different strain.

You probably recall the swine flu pandemic of 2009, when a flurry of deaths in Mexico announced the arrival of a new strain of influenza virus. Sequencing revealed that this strain, known as H1N1, had evolved by combining genetic material from four previous strains of flu. Some time earlier, three strains had combined in this way: one occurred in pigs, one in birds and one in humans. This new strain went largely unnoticed until it combined with another pig flu virus. At that point, the World Health Organization took charge and declared the virus to be a global pandemic. Governments worldwide rushed to order suitable supplies of vaccine, tailored to the new strain. In the event, H1N1 proved less dangerous than feared; by August 2010 only 15,000 people had died from the virus – far less than the millions trumpeted by the tabloid press, and less than the numbers typically killed by ordinary seasonal flu – and the response has since been criticised as overkill. But H1N1 was unusual: it had worse effects on the young than on older people who had previously been exposed to a related strain and built up immunity. It is unclear whether the authorities overreacted, or whether most of us got lucky.

• • • • • • • • • • • •

Sequence space uses multidimensional geometry as a descriptive framework, and similar ideas could be used without the associated geometric language. Twarock's work on viruses involves a deeper use of multidimensional geometry: employing detailed theorems about the intricate geometry of spaces with four or more dimensions to understand the structure of viruses.

In a series of papers, especially one published in 2004, Twarock developed a more general version of the Caspar–Klug theory, applicable to the polyoma virus and other exceptions.[9] This approach, known as viral tiling theory, allows the capsomers to be arranged in more general ways than the 'pool ball' hexagonal lattice. In particular, the pentamers need not lie at the vertices of the underlying icosahedron. Viral tiling theory is not straightforward, because regular pentagons do not tile the plane – if you try to fit them together to cover the plane, they either overlap or leave gaps – and crystallographic lattices in two and three dimensions cannot have fivefold rotational symmetry.

A key insight arose indirectly through the discovery of quasicrystal patterns, such as the famous Penrose tilings (see Figure 31). In patterns of this kind, the tiles fit together according to specific mathematical rules, but they do not form a lattice, that is, a pattern that repeats the same arrangement over and over again like a wallpaper pattern. A Penrose tiling covers the infinite plane without gaps or overlaps, using two types of tile. The resulting patterns incorporate the fivefold symmetry of the regular pentagon. Though these are not in fact lattice patterns, they can be

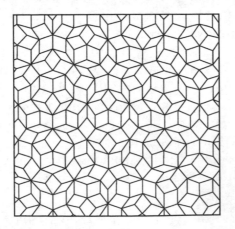

Fig 31   A Penrose tiling.

Fig 32 Non-lattice pattern in two dimensions.

understood using a mathematical trick: they are representations in 2D and 3D spaces of suitable parts of lattice patterns in spaces of higher dimension.

This idea can be understood through an analogy: constructing non-lattice patterns in 2D from lattice patterns in 3D. The lattice patterns in 2D are based on tiling by parallelograms: certain shapes of parallelogram lead to patterns with more symmetry than the rest, notably the square and hexagonal lattices. Figure 32 shows a tiling of the plane (2D) by two different shapes of tile: equilateral triangle and hexagon. This is not a lattice pattern, which among other things would use a single shape of tile, but it is very symmetric and regular nonetheless.

It turns out that we can construct the same tiling pattern from a lattice, provided we move into 3D space. The lattice required is the cubic lattice, the simplest of all 3D lattices. It is like a 3D chessboard with cubes in place of the traditional squares. The most obvious feature of a cube is its square faces, but the pattern we're considering doesn't involve squares: it is made from equilateral triangles and hexagons. Nonetheless, it is concealed within the cubic lattice pattern. In fact, if you slice this 3D pattern using a plane that is tilted to run through the midpoints of three adjacent edges of one of the constituent cubes, you get exactly the required 2D pattern.

Here, we obtain the more complex lower-dimensional pattern by taking a slice through a simpler 3D pattern. Another strategy is also available: to project parts of the 3D pattern onto a suitable plane, much as a movie image is projected from the film onto the screen. Better still, both sections and projections can be used. The details don't greatly matter: the mathematical advantage is that we can understand complex patterns in terms of simpler ones. The complexity of Penrose tilings, for example, can be simplified in this way. The price we pay is having to extend the pattern into a higher-dimensional space, but as far as mathematicians are concerned, this is harmless, easy and standard. It may sound a big deal to anyone not used to this trick, especially when the spaces involved are 4D or 5D, but algebraically it makes good sense, and that justifies using it, as well as providing a way to do the sums.

Twarock wondered whether it was possible to use this trick on icosahedral viruses, which are non-lattice patterns in 3D space. That means going to 4D space, at the very least – perhaps more. The powerful mathematics of symmetry groups can be brought to bear, and it shows that the smallest dimension of a lattice with icosahedral symmetry is not 4, but 6.

Although 6D spaces may seem irrelevant to biology, it is worth remembering that the ingredients of a mathematical description often lack direct physical counterparts. For example, the motion of the Earth and Moon around the Sun is most naturally represented as a dynamical system in 18D space, with 3 position and 3 velocity coordinates for each body, even though any configuration of the bodies lies in ordinary 3D space. So the geometry of lattices in 'unphysical' spaces of high dimension should be considered as a useful technique for determining 3D non-lattice patterns with special features, not as a literal description of a physical process.

The icosahedral group belongs to an important class of symmetry groups known as Coxeter groups after the geometer H.S. M. ('Donald') Coxeter,[10] which are higher-dimensional analogues of a kaleidoscope. Working with Tom Keef, Twarock has applied this class of groups to the structure of icosahedral viruses in terms of a 6D lattice with icosahedral symmetry known as D_6. By adapting the methods from Penrose tilings, they have constructed a class of possible virus structures defined as projections from the 6D lattice D_6 into 3D space.

The projection of the entire D_6 lattice would fill space densely with points, but the same issue arises for Penrose tilings, and the answer is to consider only a subset of D_6, a slice with non-zero thickness, so to speak, and to project only that part of D_6. This 'cut and project' technique yields all the pseudo-icosahedra, but also

additional structures with more than 12 pentamers. In particular
the structures of polyoma virus, Simian virus 40, and bacteriophage
HK97 are now accounted for.

The mathematical techniques employed here range from well-
established ideas in group theory and crystallography, through
more modern contributions such as Coxeter groups, to recent
innovations inspired by Penrose tilings. The resulting structures
have definite biological interest. One way to attack a virus is to
interfere with its assembly process, and the geometry of the fully
assembled virus provides clues about potential weak points in this
process. Viral tiling theory goes beyond the Caspar–Klug approach
by allowing different types of bonding among capsomers, so it
represents the actual molecular configurations more closely. It also
opens up new ways of thinking about tubular malformations, where
the virus assembles into a tube rather than an approximate sphere.
For example, if changes in the chemical environment of an
assembling virus can cause it to form into a tube (a non-infectious
form) rather than an infectious icosahedron – which seems
plausible – then it might be possible to interfere with virus
replication.

Other applications include cross-linking structures, such as exist
in bacteriophage HK97, in which additional chemical bonds occur
between adjacent capsomers; physical properties of capsids, which
could suggest new ways to destroy the virus; and the way the
genome is packaged within the virus. Research into this intriguing
area continues. But what is known to date fully justifies the view
that abstract geometry in higher dimensions can tell us a lot of
useful things about real viruses in three dimensions.

11 Hidden Wiring

Compared with most other animals, we have unusually large brains. We're very proud of our brains, because they are the source of human intelligence, which by any reasonable definition is greater than that of almost all other creatures (though I do wonder about dolphins). However, intelligence can't be just a matter of brain size, absolute or relative. Some animals have bigger and heavier brains than we do, and some animals have brains whose weight, in proportion to that of their bodies, is greater than ours. So brain size or weight alone seems not to imply intelligence, and neither does the relative brain size or weight (see Figure 33).

In fact, our big brains may not be as unusual as has previously

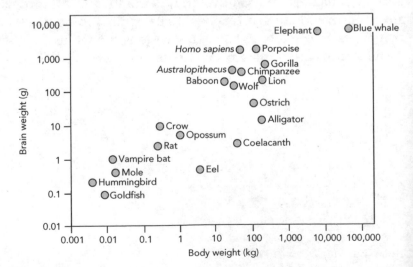

Fig 33 How brain weight relates to body weight.

been assumed. Suzana Herculano-Houzel and colleagues at the Laboratory of Comparative Neuroanatomy in Rio de Janeiro have analysed brain size in numerous species, finding that our brains are about the size you would expect in a large primate.[1] As far as brain size goes, we are no more than scaled-up monkeys.

However, monkeys are smart. A lot smarter than many creatures with brains the same size. In many types of animal, larger brains also have larger nerve cells, but in primates, nerve cells remain the same size no matter how big the brain. So a monkey brain has a lot more nerve cells than, say, a rodent brain of similar physical size.

What matters is not how big a brain you have, but what it can do and how you use it.

In ancient times, the function of the brain was a mystery. When the Egyptians mummified their pharaohs, they carefully removed the liver, lungs, kidneys and intestines, putting them in so-called canopic jars, so that the king would be able to use these vital organs in the afterlife. But they scraped out the brain by opening a hole into the skull through the back of the nose, stirred the brain until it turned to mush, drained it out and threw it away. They clearly thought that the king would not need his brain in the afterlife, and their reason was that it didn't seem to do anything.

On the other hand, like all cultures of the period, they knew that if someone's head was caved in with a club, in battle, then they would die. One of the favourite depictions on temple walls was a 'smiting scene' in which the king clubbed his enemies with a mace. So they presumably realised that you needed an intact *head* to survive, but discounted the brain because it didn't appear to have any useful function. It was just padding for the head.

The human brain is a very complicated organ, made from nerve cells: special types of cell that link together into chains and networks and send signals to one another. A typical human brain contains about 100 billion (10^{11}) nerve cells, which form up to a quadrillion (10^{15}) connections. If some recent suggestions are correct, another type of cell, called a glial cell, also takes part in the brain's processing activity, and there are at least as many of those as nerve cells – some say ten times as many.

Nerve cells, also known as neurones or neurons, do not occur just in brains. They pervade bodies as well, forming a dense network that transmits signals from the brain to muscles and other organs, and receives signals from the senses – sight, hearing, touch, and so on. Nerve cells are the hidden wiring that makes bodies work. Even lowly creatures such as insects have complicated networks of neurons. The nematode *Caenorhabditis elegans* is a tiny worm, much studied because it always has the same number of cells in the same layout: 959 in the adult hermaphrodite, 1,031 in the adult male. One-third of them are nerve cells.

• • • • • • • • • • • •

Even a single nerve cell is complex. But what makes nerve cells so powerful as signalling and data-processing devices is their ability to network. The main body of a nerve cell has many tiny protuberances, known as dendrites, which receive incoming signals (see Figure 34). The nerve cell also transmits outgoing signals, which travel along its axon, the biological equivalent of a wire. The axon can branch at its far end, so the same signal can be fed to many different recipients. The signals are electrical, as in modern communications, and they use very small voltages; they are generated, distributed and acted upon through chemical reactions. The simplest signal is a short, sharp pulse of electrical activity, but

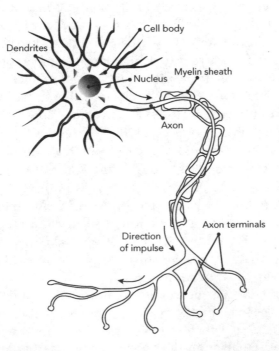

Fig 34 Nerve cell.

nerve cells can also produce signals that oscillate, or occur in short bursts, or behave in more complicated ways.

By connecting axons to dendrites, nerve cells can form structured networks. The mathematics of such networks, which we'll get to in due course, shows that the resulting dynamics can be far more complicated than anything the component nerve cells can do on their own, just as a computer can do things that a transistor cannot. However, even a single nerve cell is a complicated thing to describe and model mathematically.

Scattered through the body are many different specialised networks of neurons, which cause muscles to contract or detect and process sensory data. Networks with just a few neurons can perform sophisticated tasks. Large networks of a few hundred are already too complex to understand in detail, and they can do many things that small networks can't. A network of 100 billion neurons – a brain – poses a serious challenge for biologists and mathematicians alike. In fact, there is no real prospect of gaining a complete understanding: the brain is too complex. Nonetheless, a lot of progress is being made, because to some extent the brain has a modular structure, and we can study individual modules, which are simpler.

. . . .

The simplest such module is a single nerve cell. If you can't understand how a single nerve cell works, you won't get far with an entire brain. One of the first significant applications of mathematics to biology occurred in neuroscience, the study of the nervous system, in 1952. The problem was the transmission of individual pulses of electricity along a nerve axon – the basis of the signals that are sent from one neuron to another. The Cambridge biophysicists Alan Hodgkin and Andrew Huxley developed a mathematical model for this process, now called the Hodgkin–Huxley equations.[2] Their model describes how an axon responds to an incoming signal received by the nerve cell. They were awarded a Nobel prize for this work.

Hodgkin and Huxley started from a physicist's model of the nerve axon, treating it as a poorly insulated cable transmitting electricity. The insulation is poor because some of the atoms that take part in the associated chemical reactions can leak out. More precisely, what leaks is ions: charged atomic nuclei. The main vehicles of voltage leakage are sodium ions and potassium ions, but others, especially calcium, also play a role. So Hodgkin and Huxley wrote down the standard mathematical equation for electricity passing along a cable, and modified it to take account of three types of leakage: loss of sodium ions, potassium ions, and all other ions (mainly calcium). The Hodgkin–Huxley equations state that the electrical current in the cable is proportional to the rate of change of the voltage (this is Ohm's law, a simple and basic piece of the physics of electricity), together with additional terms that account for the three types of leakage.[3]

The actual equations are messy, because of the complicated form of the leakage terms, and can't be solved by a formula, so Hodgkin and Huxley did what all scientists and mathematicians do in such circumstances: they solved the equations numerically. That is, they calculated very good approximations to the solutions. There were already excellent, long-established methods for doing that, and yet another branch of mathematics, numerical analysis, entered the story. They did not have a computer; hardly anyone did in those days, and those that existed were the size of a small house. So they carried out the calculations by hand using a mechanical calculator. The result was that a voltage spike should travel along the axon (see Figure 35). With specific values for various data, derived from experiments, they calculated how fast the spike travelled. Their figure of 18.8 metres per second compared well with

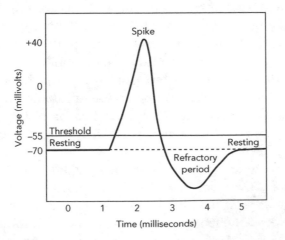

Fig 35   Voltage spike predicted by the Hodgkin–Huxley model.

the observed value of 21.2 metres per second, and the calculated profile of the spike was in good agreement with experiment.

The spike has some important features, which gave some insight into how a nerve cell works. The incoming signal has to be greater than a particular threshold value before the nerve cell fires and triggers a spike. This prevents spurious outgoing signals being triggered by low-level random noise. If the incoming signal is below the threshold, the voltage in the axon bumps up slightly but then dies away again. If it is above the threshold, the dynamics of the nerve cell causes the voltage to increase sharply, and it then dies down even more sharply; these two changes create the spike. There is then a short 'refractory period' during which time the nerve cell does not respond to any incoming signal. This keeps the spikes separate (and spiky). After that, the cell is back at rest and ready to respond to the next signal it receives.

. . . .

Today there are many mathematical models of a single nerve cell or axon. Some sacrifice realism for simplicity, and are even simpler than the Hodgkin–Huxley equations; others aim at greater realism, which automatically makes them more complicated. As always, there is a trade-off: the more features of the real world you put into the model, the harder it becomes to work out what it can do. The goal, not always attainable, is to retain the features that matter and discard everything that's irrelevant.

One of the simplest models yielded a valuable insight: the nerve axon is an excitable medium. It responds to a small input by amplifying it; then it temporarily switches off the amplification process, so that the resulting signal cuts off at some finite value instead of rising indefinitely. The model concerned derives from Richard Fitzhugh's work at the National Institutes of Health at Bethesda in Maryland in the early 1960s, and is known as the FitzHugh–Nagumo equations.[4] FitzHugh made conscious mathematical simplifications to the Hodgkin–Huxley equations, combining the roles of the ionic pathways into a single variable. The other key variable is the voltage. So the FitzHugh–Nagumo equations are a two-variable system, and we can represent those variables as the two coordinates of a plane. In short, we can draw pictures.

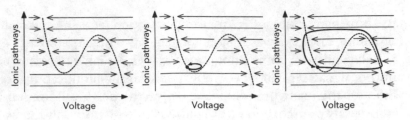

Fig 36 Excitability in the FitzHugh–Nagumo equations. *Left*: Direction in which the voltage changes. *Middle*: A small kick dies out and the state returns to rest (black dot). *Right*: A large kick triggers a voltage spike followed by a slow return to rest.

Figure 36 shows the most important feature of the FitzHugh–Nagumo equations: excitability. The left-hand picture shows a wiggly curve in the plane of the two state variables: this is the curve on which the voltage would not change as time passes if the ionic currents were fixed. The arrows show the direction in which the voltage changes as time passes. The middle picture shows the effect of a small disturbance, or 'kick', produced by an incoming signal. The state is disturbed from rest, but fails to cross the dotted curve, so it makes a small excursion and returns rapidly to its resting state. The right-hand picture shows the effect of a larger kick. Again the state is disturbed from rest, but now it crosses the dotted curve, so it makes a *large* excursion and eventually returns, slowly, to its resting state. This property is called excitability, and it is one of the central mathematical features of both the Hodgkin–Huxley and FitzHugh–Nagumo models. Excitability is what allows the neuron to generate a large voltage spike when given a small *but not too small* kick, and then return to rest reliably, even if a further signal comes in which might interfere with that process.

This sequence of events shows how a neuron obeying the FitzHugh–Nagumo equations can generate a single, isolated voltage spike. Real nerve cells do this, but they also generate long trains of pulses – they oscillate. Similar pictures reveal that the FitzHugh–Nagumo model can also produce oscillations.

This is just the simplest model for nerve cell dynamics. There are many others, and which is appropriate depends on the question being answered. The more powerful your computer, the more 'realistic' the equations can be made. But if the model gets too complicated, it often yields little insight beyond 'it does so-and-so because the computer says so'. For some questions, a simpler but less realistic model may be better. This is the art of mathematical modelling, and it is more of an art than a science.

Excitability is one of the reasons why nerve cells can send one another signals. In fact, it is why signals can be produced to begin with. But the really interesting behaviour arises when several cells send signals to one another. Nerve cells in real animals form complex networks. Mathematical biologists are just beginning to grasp the amazing power of networks. A network of relatively simple components, communicating via little more than series of spikes, can do extraordinary things. In fact, they can probably do everything our brains can do, as far as manipulating sensory inputs and generating outputs to the body are concerned. This seems likely because the brain is a very, very complex network of nerve cells. And there are good reasons to think that most of the brain's astonishing abilities are consequences of the network architecture.

. . . .

One of the biological topics that I've worked on myself provides a nice example of networks of nerve cells. The topic is animal locomotion: how animals move using their legs, what patterns they use, and how those patterns are produced. This is a huge subject in its own right, with a fascinating history, but I can only scratch the surface here.

In July 1985, along with two other mathematicians and a physicist, I was travelling through redwood and sequoia forests down the coast of California in a Mini. We were on the way home from a mathematics conference in Arcata, a small town about 200 miles north of San Francisco. To pass the time when we weren't hopping out of the car to look at giant trees, Marty Golubitsky (one of the mathematicians) and I started thinking about the patterns that form when you hook a lot of identical units together into a ring.

We'd already sorted out a general method for approaching this sort of question. It predicted that rings of this kind should generate travelling waves, in which successive units round the ring do exactly the same thing, but with a time delay. The simplest example is to hold your arms out sideways and let them dangle from the elbows. Now let the dangling bits swing to and fro. Typically, they either synchronise, with both arms moving in the same way, or they anti-synchronise, with the left arm doing the exact opposite of the right.

Similar patterns arise when there are more than two components. For instance, if you hook four components together in a square, with each one connected to the next, then you can get patterns in which all four components oscillate periodically, but there is a time difference of one-quarter of the period between each component and the next. It's like four people all singing the same four-beat bar of music over and over again, but one starts with the first note, the next starts with the second note, the next with the third note and the last one with the fourth note.

Later, we realised that similar patterns occur in animal locomotion. Animals move in a variety of patterns, called gaits, repeating the same sequence of movements over and over again in a series of 'gait cycles'. A horse, for example, can walk, trot, canter or gallop – four distinct gaits, each with its own characteristic pattern. In the walk, the legs move in turn, and each hits the ground at successive quarters of the gait cycle. So the sequence goes left back, left front, right back, right front, all equally spaced, over and over again. The trot is similar, but one diagonal pair of legs hits the ground first, and the other pair does so half a gait cycle later. So in both gaits all four legs do essentially the same thing, but with specific differences in the timing.

There are ten or twelve common patterns in animal gaits, and dozens of uncommon ones. Some animals, such as the horse, use several different gaits. Others use only one, other than standing still. Four of the most common gaits are illustrated in Figure 37.

Jim Collins, a biomechanicist (someone who applies mechanics to biology, especially medicine) at Boston University's medical institute heard of our ideas, and he told us that gaits are thought to

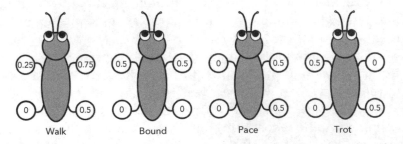

Walk   Bound   Pace   Trot

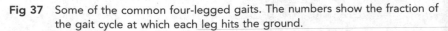

Fig 37 Some of the common four-legged gaits. The numbers show the fraction of the gait cycle at which each leg hits the ground.

be produced by a relatively simple circuit in the animal's nervous system, known as a central pattern generator (CPG). This is located not in the brain, but in the spine, and it sets the basic rhythms, the patterns, for the movement of the muscles that actually causes the animal to walk, trot, canter or gallop.

No one had actually seen a CPG at this time. Their existence was inferred indirectly, and in some quarters they were a bit controversial, but the evidence was quite strong. However, the exact network of connections among the nerve cells was unknown. So we did the best we could and worked out the most plausible patterns, assuming various more or less natural structures for the CPG. We started with quadrupeds, and went on to apply similar ideas to six-legged creatures: insects.

There had already been a lot of work done on the mechanics of legged locomotion. Our approach was more abstract, trying to infer the structure of a hypothetical CPG from the patterns observed in the legs. But it had an interesting consequence: it revealed that the same network of nerve cells, operating under different conditions, is capable of generating *all* the most symmetric quadruped gaits – and, in more subtle circumstances, less symmetric ones like the canter and the gallop as well.

None of the networks that we proposed was completely satisfactory, for various technical reasons. Discussions with Golubitsky and the Ontario-based mathematician Luciano Buono led to the insight that any workable CPG for quadrupeds must have at least two units per leg: one to work the muscles that flex the leg and the other to work the muscles that extend it. So the most natural CPG for quadrupeds should have eight units (Figure 38, see over).[5]

This network can generate all the basic four-legged gaits. Along with the standard gait patterns, it predicted one that we'd not encountered before. In this gait, which we named the jump, the two rear legs hit the ground together, and then the two front legs hit the ground together a quarter of the way through the gait cycle. If it had been halfway through the cycle, this gait would have been a standard one, the bound. Dogs bound when running fast, for example. But one-quarter of the way through the gait cycle was a real puzzle, especially since no legs hit the ground halfway through or three-quarters of the way through. It was as though the animal were somehow suspended in mid-air.

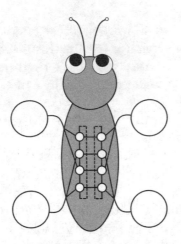

Fig 38 Predicted architecture of quadruped CPG. Two rings of four identical modules are linked left–right. Two modules connect to each leg, determining the timing of two muscle groups. The picture is schematic and there may be many more connections in the CPG, but having the same symmetry.

We came to this conclusion late one afternoon. The Houston Livestock Show and Rodeo was in town, and we had seats booked for that evening. So we went to the Astrodome, and watched cattle being roped and buggies being raced. Then came the bucking broncos. The horses were trying to throw the riders off their backs, and the riders were trying to stay on for as long as they could – which was often just a few seconds. Suddenly Golubitsky and I looked at each other and started counting ... The horse that we were watching was jumping into the air, both back feet giving it a push, then both front, then hanging in the air ...

It looked very much as though the difference in timing was one-quarter of the full gait cycle. The Exxon Replay, a television recording in slow motion, of that precise horse confirmed this. We had found our missing gait.

Later we discovered that two other animals, the rat and the Asia Minor gerbil, also employ this unusual gait. We found several other features of real gaits that were predicted by our eight-unit CPG or natural generalisations, including one displayed by centipedes. Of course none of this proved that our theory was correct, but it did mean that it passed several tests that might prove it wrong.

More recently, Golubitsky and the Portuguese mathematician Carla Pinto have applied the same idea to a four-unit network for biped gaits – two legs, like us.[6] They find ten gait patterns, eight of which correspond to known gaits in bipeds. They include the walk, run, hop and skip.

. . . .

Another fascinating network of neurons is definitely found in a real animal, rather than being a mathematician's pipe-dream. It is the CPG for the heartbeat of the medicinal leech *Hirudo medicinalis*.

Leeches are slug-like creatures that suck blood. They were widely used in ancient medicine to 'balance the body's humours' when there was deemed to be an excess of blood; the earliest record of their use goes back to 200 BC. The use of leeches died out during the nineteenth century, but by the beginning of the twenty-first they were back in vogue, though for more scientific reasons and in a more limited realm. The saliva of leeches contains a molecule called hirudin, which keeps blood flowing freely while the animal sucks it from its victim. Hirudin helps to stop the patient's blood coagulating during microsurgery.

The medicinal leech poses a curious problem for mathematicians interested in networks of neurons: its heartbeat follows a very strange pattern. The heart of a leech consists of two rows of chambers, with about 10 to 15 in each row, depending on the species. The pattern goes like this. For a while, all the chambers on the left beat in synchrony – at the same moment. While this is happening, the chambers on the right beat in sequence, one after the other, from back to front. After between 20 and 40 beats, the two sides swap roles. Then they swap again, and so on.

Nobody really knows *why* the leech heart beats in this manner, but the blood pressure is high when the chambers beat in sequence, and much lower when they beat synchronously, so the need to avoid persistent high pressure (or indeed persistent low pressure) may be part of the reason. We have a much more complete explanation of *how* it does it. The strange switching between two patterns is driven by the creature's nervous system, and is a natural feature of network dynamics (Figure 39, see over).

Ronald Calabrese and colleagues at Emory University in Atlanta, Georgia have investigated the heartbeat of the leech in an extensive series of papers.[7] They traced the dynamics to a CPG: a network of nerve cells located in most of the leech's 21 segments. In each

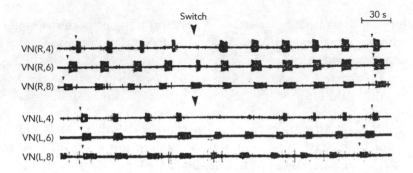

Fig 39 Recordings from various vascular nerves of a leech reveal short bursts of electrical activity. Here the 4th, 6th and 8th nerve cells on the right side of the leech (the top three rows of signals) are bursting in sequence before the time marked with an arrow, but in synchrony after that. The corresponding nerve cells on the left side of the leech (the bottom three rows) initially burst in synchrony, but switch to bursting in sequence.

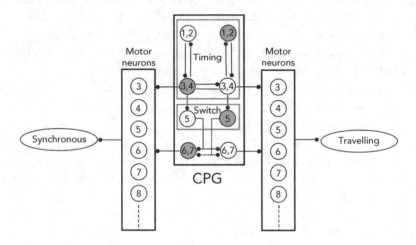

Fig 40 CPG for the leech heartbeat.

segment there is a pair of motor neurons, one on the left and the other on the right, which make the heart muscles contract (see Figure 40). There is also a pair of 'interneurons', which help to generate the pattern of nerve impulses needed to control the heartbeat. The interneurons in the third and fourth segments are wired up to produce a regularly pulsing timing pattern; in effect they form a clock whose regular ticking influences what the other neurons do. The wiring in the fifth, sixth and seventh segments directs these signals to the heart motor neurons on left and right sides of the leech, and modifies the signals so that on one side the effect is a sequential wave of contractions, but on the other all muscles contract simultaneously. The dynamics of these three segments switches regularly from this pattern to its mirror image.

Calabrese's group originally focused mainly on these timing circuits, and did not investigate in any detail how the timing signals are transmitted to segments further along the leech. In 2004 Buono and Antonio Palacios, of the Nonlinear Dynamical Systems Group at San Diego State University, employed techniques from dynamical systems with symmetry to model this transmission process.[8] They modelled the network that transmits the signals as a chain of neurons whose ends are linked to form a closed loop. Symmetric dynamics tells us that there are two common patterns of periodic oscillation in closed loops: sequential and synchronous. The relations between the two patterns arise through a so-called mode interaction, when the parameters of the network connections cause both patterns to arise simultaneously.

In earlier work by Buono and Golubitsky, mode interactions of this type were invoked to explain the less-symmetric gaits of quadrupeds, such as the canter and gallop of a horse. So the cantering horse and the heartbeat of the leech fall into the same category of phenomena, both mathematically and biologically.

More recently, Calabrese's group has devised more detailed models, with an emphasis on the structure of the signals between nerve cells, which occur in short bursts (as shown in Figure 39). The role of bursting seems to be central to this and many similar problems in neuroscience, so mathematicians have developed equations for bursting neurons.

. . . .

Networks of neurons are involved in perception as well as motion.

In 1913 the New York neurologists Alwyn Knauer and William Maloney published a report in the *Journal of Nervous and Mental Disease* on the effects of the drug mescaline, a psychedelic alkaloid found in cacti, notably peyote, which grows in desert areas of Central America. Their subjects reported striking visual hallucinations:

> Immediately before my open eyes are a vast number of rings, apparently made of extremely fine steel wire, all constantly rotating in the direction of the hands of a clock; these circles

are concentrically arranged, the innermost being infinitely small, almost point-like, the outermost being about a meter and a half in diameter. The spaces between the wires seem brighter than the wires themselves ... The center seems to recede into the depth of the room, leaving the periphery stationary, till the whole assumes the form of a deep tunnel of wire rings ... The wires are now flattening into bands or ribbons, with a suggestion of transverse striation, and colored a gorgeous ultramarine blue, which passes in places into an intense sea green. These bands move rhythmically, in a wavy upward direction, suggesting a slow endless procession of small mosaics, ascending the wall in single files ... Now in a moment, high above me, is a dome of the most beautiful mosaics ... Circles are now developing upon it; the circles are becoming sharp and elongated ... now all sorts of curious angles are forming, and mathematical figures are chasing each other wildly across the roof.

Similar effects can be seen by closing your eyes and pressing on the eyeballs with your thumbs, so a first guess might be that the drug affects the eyes. Actually, it affects the brain, creating signals that the brain's visual system *interprets* as images seen by the eye (see Figure 41). These patterns offer important insights into the structure of the visual system, thanks to a mixture of experiments and mathematical analysis.

A large part of the brain is devoted to the visual system. Biologists have been studying the neuroscience of vision for many years, and have learned a lot about it. But there's a lot that we still don't understand. At first sight, you may wonder what the problem is. Isn't the eye basically a camera – a pinhole camera with the pupil as a pinhole, a lens to improve the focus and a retina to receive the image? The main problem is that the brain doesn't just

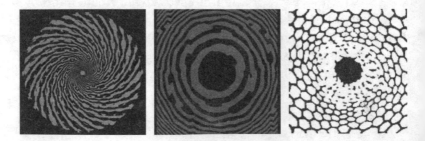

Fig 41   Hallucination patterns. *Left*: Spiral. *Middle*: Tunnel. *Right*: Honeycomb.

passively 'take a photograph' of what is out there. It provides automatic understanding of what the eye is seeing. The brain's visual system processes the image, working out what objects are being viewed, where they are relative to one another, even decorating them with the vivid colours that we perceive. We look out of the window and instantly 'see' a man walking past with his dog. But the dog is passing behind a lamppost, and half the man is hidden from view by a hedge; the man's body is mostly covered by a coat and his face is obscured by the hood. Moreover, the image appears to be three-dimensional: we know, without thinking, which parts are in front of which.

It has proved almost impossible to teach a computer to analyse such a scene and recognise the main objects in it, let alone work out what they are doing. Yet the visual system achieves this, and more, in real time, with a moving image. With the help of some clever biochemistry and difficult experiments, mathematical biologists are starting to make inroads into the structure and function of the visual system. And what we currently understand shows that evolution has been very clever indeed.

Images received by the retina of each eye pass along the corresponding optic nerve (actually a huge bundle of nerve fibres) to an area of the brain known as the visual cortex. This lies on the surface of the brain, just under the skull, and if you were to lift off the top of the skull you would see that it has the familiar convoluted shape, a bit like a cauliflower. If you flatten out the convolutions, the visual cortex turns out to be made from a number of layers of nerve cells, placed on top of one another. The top layer, known as V1, starts by representing the image as an array of on/off signals in the corresponding nerve cells. In experiments, it proved possible to work out what a cat was looking at by using voltage-sensitive dyes to make the on/off pattern visible. If the cat was looking at a square, say, then a distorted square appeared on the V1 layer.

Each layer of the cortex is connected to those above and below, and this wiring is done in such a way that successive layers extract different information from the basic image formed on the top layer. The next layer down, for example, sorts out the boundaries between different features of the image – where, say, the dog's body appears to be cut by the edges of the lamppost, and where the man's hood ends and the house behind it begins. It also works out the orientations of these boundaries. Then the next layer can compare directions and locate such things as the dog's eyes, which are point-like, so the edges change direction very rapidly. Presumably, many layers down, is a nerve cell or a network of them that makes the deduction 'dog'; after that comes the recognition that the dog is a Labrador retriever, until at some level you realise that it's Mr Brown taking Bonzo for his evening walk.

There are useful mathematical models of the first few stages of this process, which have a very geometric feel to them. These models show how information flows back up through the layers as well as down into the depths, priming the visual system actively to look for specific features. This is how our eyes track the dog as parts of it vanish behind the lamppost and reappear. It 'knows' that they are going to do that, and anticipates it. It is possible to devise experiments that make use of the visual system's ability to anticipate what it expects to see, by tricking it into seeing something else. It is also possible to alter the brain's perceptions using drugs. Recent discoveries about the V1 layer of the visual cortex depend, to some extent, on the patterns that arise when volunteers take hallucinatory drugs such as LSD.

Legal experimental volunteers and illicit users of these drugs report seeing a variety of strange, geometric patterns. It can be shown that the patterns do not originate in the eye, even though they are 'seen' to be part of the image that the eye is receiving. Instead, they are artefacts caused by changes in the way the nerve cells in the cortex function. Anything that the brain's networks perceive is automatically interpreted *as though* it had come from an external image. We can't stop that happening; it is how we see everything. What we fondly imagine to be the external world is actually a representation held in our own heads. This is the main reason why the visual system can be tricked into seeing things that aren't there. On the whole, though, we can trust our visual system, which evolved to perceive things that *are* there. Except when it is presented with carefully contrived images that create illusions, or drugs that cause hallucinations.

Around 1970, Jack Cowan, a biomathematician now at Chicago, started to use hallucination patterns to unravel the structure of V1. His first discovery provided strong evidence that hallucination patterns arise in the brain, not the eye. The range of reported hallucination patterns is huge, but most of them are variations on a basic theme: spirals. Some are concentric circles, and a circle is what happens to a spiral when it is wound so tightly that the spacing between successive turns becomes zero. Others are radial spokes, another special case when the turns of the spiral grow so rapidly that the spacing effectively becomes infinite. Sometimes the spiral pattern is decorated with hexagons, like a honeycomb; sometimes it is covered in a chequered pattern of lozenges, like Elizabethan window panes. Sometimes it is too bizarre to describe meaningfully. But spirals dominate, and they are of the type that mathematicians call logarithmic. Spirals and spokes often rotate, and concentric circles can spread like the ripples on a pond when you drop a stone into it.

To form a logarithmic spiral, imagine a spoke that rotates at constant speed, and a point that moves outwards along the spoke. The combination of these two motions creates a spiral, but the exact shape of the spiral depends on how the point moves along the spoke. For example, if it travels at a fixed speed, we get an Archimedean spiral, in which successive turns are equally spaced. If the speed grows exponentially, so that (say) each turn is twice as far out as the previous one, we get a logarithmic spiral (see Figure 42).

In 1972, Cowan and his collaborator Hugh Wilson wrote down mathematical equations, now called the Wilson–Cowan equations, which describe how a large number of interconnected neurons interact with one another.[9] In 1979 Cowan and Bard Ermentrout used these equations to model how waves in the visual cortex propagate, and deduced that the complex spiral patterns reported by experimental subjects can be explained in terms of much simpler patterns of electrical and chemical activity in the top layer of the cortex. The predominantly spiral nature of the hallucinations is a

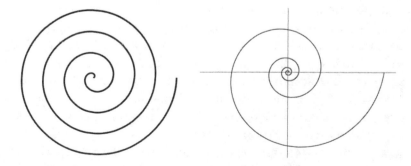

Fig 42 *Left*: Archimedean spiral. *Right*: Logarithmic spiral.

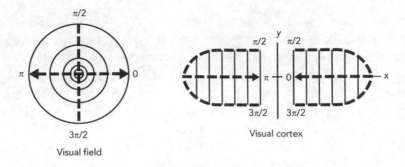

Fig 43 Retino-cortical map.

useful clue, because physiologists have figured out how the image sent to the cortex by the retina compares to the one that the cortex actually receives. The retina is circular, but the V1 layer, when unfolded, is roughly rectangular. The image on the cortex is a distorted version of the image on the retina. The way it is distorted can be described by a mathematical formula which has the effect of converting radial lines on the retina into parallel lines on the cortex, and concentric circles on the retina into another set of parallel lines, at right angles to the first. However, just to keep us on our toes, this happens in two different regions of the cortex. In reality they are joined like an hourglass, as in Figure 44, but mathematically it is often convenient to swap the two ends to make an oval shape, as in Figure 43.

The most significant feature of this 'retino-cortical map' is that it turns logarithmic spirals on the retina into parallel lines in the cortex. The visual system automatically makes us 'see' anything detected by the cortex, whether or not it originates in the eye, so we will 'see' spirals if something creates a pattern of parallel stripes on the cortex. Psychotropic drugs cause waves of electrical activity, and the simplest wave pattern is stripes.

If the stripes move, forming travelling waves like ocean waves coming up a beach, then the geometry of the retino-cortical map implies that the spirals will appear to rotate. Radial spokes are a special case of spirals, and correspond to horizontal stripes on the cortex (relative to the conventional orientation of the cortex). As the waves travel across the cortex, the spokes rotate. Concentric circles similarly correspond to vertical stripes; this time, as the waves travel across the cortex, the circles appear to expand from their centre (see Figure 44).

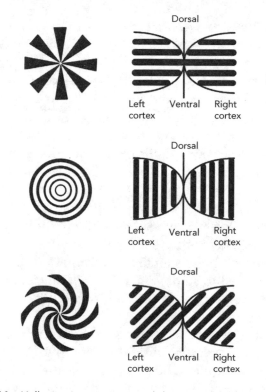

Fig 44 Hallucination patterns and the corresponding waves in the cortex.

This already offers a big clue about the hallucination patterns: they make much more sense if they are caused by simple striped waves, moving across the cortex. And those waves are patterns of electrical and chemical activity, caused by the drug. The other more elaborate patterns sometimes reported have a similar explanation, involving different behaviour on the cortex. The spiral honeycombs, for instance, correspond to straightforward honeycomb patterns on the cortex.

The equations that Cowan and Ermentrout devised constitute what mathematicians call a continuum model. That is, it models the very fine but discrete network of real neurons by a continuous distribution of infinitesimal neurons. The cortex becomes a plane, and the neurons reduce to points. This kind of transition from discrete reality to a continuous model has become a standard strategy when applying mathematics to the real world, because it permits the application of differential equations, a very powerful tool. Historically, the first areas of science to be treated this way were the movement of fluids, the transfer of heat and the bending of elastic materials. In all three cases, the real system is composed of discrete atoms, which though very small are indivisible, whereas the mathematical model is infinitely divisible. Experience shows that continuum models are very effective provided the discrete components are much bigger than the effects being described. Although neurons are much larger than atoms, they are considerably smaller than the wavelengths of electrical waves in the cortex, so it is reasonable to hope that a continuum model might be worth pursuing.

A further simplification replaces the oval shape of the flattened-out cortex by an infinite plane. The modelling assumption here is that the edges of the cortex do not have a significant effect on waves in its interior. With all these simplifying assumptions in place, standard methods used to study pattern formation can be applied to classify the wave patterns in the cortex, and hence the hallucination patterns.

Initial attempts to analyse travelling waves in the cortex using these methods scored some successes, but didn't always get the details right. When new experimental methods revealed the 'wiring diagram' for neurons in the cortex, it became clear why this was happening, and suggested a slightly different model.

Cells in V1 do not just map out the image, like a collection of pixels on a TV screen. They also sense the directions of edges in the image. So the state of a cell is not just the brightness of the corresponding point in the image (I'm ignoring colour here and thinking about just one eye) but also the local *orientation* of any lines in the image. Mathematically, each point in the plane has to be replaced by a circle of orientations. The resulting shape does not correspond naturally to a shape in our familiar three-dimensional space, but we don't need to visualise it in order to do the sums.

Experiments show that the neurons in V1 are arranged in small patches, and within each patch the neurons are especially sensitive to a particular orientation. Neighbouring patches respond to nearby orientations, so that (say) one patch might respond to vertical edges, and a neighbouring patch might respond to edges tilted 30° to the right of vertical. Most of the connections among these neurons fall into two types. Within any given patch, we find short-range inhibitory connections, meaning that incoming signals suppress activity in the nerve cell instead of stimulating it. But there are also long-range excitatory connections between distinct patches, and they do stimulate activity. Moreover, the excitatory connections are aligned along a specific direction in the cortex: the same direction that the patch itself prefers.

This pattern of connections changes the corresponding continuum model, which is not a plane's worth of points, but a plane's worth of circles. Translations and reflections behave in the usual way. However, any rotation of the plane must also rotate every circle, otherwise the wiring diagram doesn't work out properly. Once the experiments have shown how to puzzle out the correct model, the heavy machinery of the mathematics of pattern formation can be brought to bear.[10] This time, the classification of the wave patterns in the cortex, and the corresponding hallucinations, performs better.

The new mathematical model provides a convincing catalogue of hallucination patterns, and it also suggests the reason for the wiring diagram that nature has evolved for the V1 layer of the cortex. Imagine one of the patches, and suppose that it 'sees' a short segment of horizontal line. Its local inhibitory connections effectively vote for the most plausible direction for this line: the direction that receives the strongest signal wins, and all other orientations are suppressed. But now that patch sends excitatory signals to patches that are sensitive to that same orientation, in effect biasing them in favour of the same orientation. And it sends this signal only to patches that lie along the continuation of the direction that it has selected. The net result is that the direction of the line, as seen by this patch, is tentatively extended across the cortex. If this extension is correct, the next patch across will reinforce that choice of direction. However, if the next patch detects a strong signal in a different direction, then this will override the bias signal that is attempting to extend the edge.

In short, the nerve cells are wired so that they fit local bits of edge together into longer lines in the most plausible fashion. Any small gaps in these lines will be 'filled in' by the excitatory signals; however, if the line changes direction, the local inhibitory signals will confirm that this has happened and settle on a new choice. So the end result is that V1 creates a series of contours – linear outlines of the main features in the image. In 2000, John Zweck, a mathematician at the University of Maryland, and Lance Williams, a computer scientist at the University of New Mexico, used exactly the same mathematical trick to devise an effective algorithm for their work on computer vision.[11] The application was contour completion – filling in missing bits of edges in an image, for example where part of one object is hidden behind another.

Like the dog partly hidden by the lamppost.

. . . .

Neuroscience is one of the most active areas of mathematical biology. Researchers are working on a huge variety of topics: how neurons work, how they link up during development, how the brain learns, how memory works, how incoming information from the senses is interpreted. Even the more elusive aspects of the human brain, such as its relation to mind, consciousness and free will, are also under investigation. The techniques employed include dynamics, networks and statistics.

In parallel with these theoretical developments, biologists have made major advances in experimental techniques to study what the brain is doing. There are now several ways to image the activity of the brain in real time – in effect, to watch which parts of the brain are active, and how the activity passes from one region to another. But the brain is immensely complex, and at the moment it is probably better to focus on specific features of the nervous system, rather than trying to understand the whole thing in one go. Our brains are so complicated that, ironically, they may be inadequate to understand ... our brains.

12 Knots and Folds

· ·

The discovery of the genetic code, which represents amino acids in proteins as triples of DNA bases, was the first of many breakthroughs where DNA was thought of as a code, a list of symbols. But the physical form of the DNA molecule is also important, and so are the forms of the molecules that it encodes.

The same goes for proteins. To make an organism, you need more than a list of proteins: you need to get the right proteins into the right places at the right time. Just as you can't make a cake by dumping all the ingredients into a bowl and sticking it in the oven, you can't make an organism by making 100,000 proteins and hoping that they will somehow sort themselves out into an amoeba or a human being.

Only a small proportion of the genome consists of genes – codes for proteins. For a long time the rest was stigmatised as 'junk DNA', evolutionary relics with no current function, going along for the ride because it wasn't worth evolution's while to weed them out. It is now clear that at least some of that junk DNA helps to control how an organism is assembled. The rest may still be junk – but I wouldn't bet on it, given the past record.

How does this DNA control system work? The answer might require no more than the cracking of some other code, the code for instructions rather than ingredients. But again it's not that simple. One of the systems that control an organism's development uses genes (more properly, the proteins they encode) to switch other genes on and off. What matters here is the dynamics of the genetic switching network. And dynamics is a matter for mathematics, it can't just be read off from DNA codes. You might, perhaps, be able to read off which genes act on which other ones – but that information won't tell you what they all do when everything is happening at once. In the same way, knowing how temperature and humidity in the Earth's atmosphere affect each other doesn't tell you next week's weather.

The shapes of the molecules turn out to be at least as important as the sequences that determine them. The double-helix shape of DNA governs many of its most basic properties. In particular, the copying system for DNA, which cells and indeed the entire organism use to reproduce, must overcome a massive topological obstacle: the two DNA strands are twisted round each other like the strands of a rope. If you try to pull a rope apart by tugging on its strands, all you get is a hopeless tangle.

The shape of a protein is more important still. Many proteins do their jobs by binding to other proteins – sticking to them, usually temporarily, but in a controllable way. When the protein haemoglobin picks up or releases a molecule of oxygen, it changes shape. A protein is a long chain of amino acids, and it gets its shape by folding up into a compact tangle. In principle, the shape of this tangle is determined by the sequence of amino acids; in practice, it's virtually impossible to calculate the shape from the sequence. The same sequence can fold up in a gigantic number of ways, and it is generally thought that the actual shape it chooses is the one with the least energy. Finding this minimal energy shape, among the truly gigantic list of possibilities, is a bit like trying to rearrange some list of thousands of letters of the alphabet in the hope of getting a paragraph from Shakespeare. Running through all the possibilities in turn is totally impractical: the lifetime of the universe is too short.

. . . .

One of the keys to the mysteries of DNA shape is a branch of mathematics known as topology. As a well-developed area, topology has been around for little more than a century, though with hindsight a few precursors can be detected. By the 1950s it had shot to stardom, becoming one of the central pillars of pure mathematics, but its role in applications was still relatively minor. It clarified some theoretical issues in the dynamics of the Solar System, for instance. Topology is important in pure mathematics because it provides conceptual machinery to deal with any question involving *continuity*. And continuity – transforming shapes and structures without tearing them apart or breaking them into separate pieces – is a common theme in many different areas of mathematics. It is a common theme in applications, too: most physical processes are continuous. But it's not straightforward to deduce anything useful from that property, and it took a lot longer for topology to find a role in applied science.

Topology is not included in school mathematics lessons, except for a few cute but inconclusive tricks. A typical example is the Möbius band, invented independently by August Möbius and Johann Listing in 1858. Take a long strip of paper, bend it round to bring the ends together like a dog collar, twist one end through half a turn, and then glue the ends together. The resulting surface has several counter-intuitive properties: it has only one side, and only one edge, and if you cut it along the middle it does not fall apart into two separate pieces. There are a couple of practical applications of Möbius bands: conveyor belts that last twice as long before they wear out, and a method for connecting electrical wiring to a rotating object. But none of that is terribly impressive.

There is a lot more to topology, but the concepts are too abstract to be explained easily or accurately without a lot of technical background. However, the pure mathematicians' nose for an important idea has eventually been vindicated, and topological methods are being used in an increasingly broad range of real-world problems, from biology to quantum field theory. The application that I will describe here has yielded some crucial insights into the workings of DNA. The topological gadgetry that comes into play is less technical than in most other areas of applied maths – and is something we all come across almost every day.

Knots.

. . . .

Knots tend to be associated with tying parcels, boy scouts, sailing and mountaineering. Centuries of trial and error have shaped our understanding of knots: which one to use in which circumstances. Tie a pig to a square post with a clove hitch and you're in trouble; if the pig runs round the post in the right direction, the knot obligingly unties, and off goes your bacon.

The topology of knots tackles two general questions. One is to decide whether two knots are topologically the same, that is, whether each can be changed into the other by a continuous transformation. If you tie one of them in a length of string, can you twist the string to get the other knot? A special case of this question is to determine when an apparently complicated knot is really unknotted (see Figure 45). The second topological question is much more ambitious: whether we can classify all possible knots. There are infinitely many knots, and even the simpler ones, with few crossings, provide a rich variety of types.

Any knot tied in a length of string can be untied by reversing the process; then you can retie the string into any other knot that you wish. To get a sensible question, we have to do something to stop the knot escaping off the end of the string. The time-honoured solution in topology is to glue the ends of the string into a closed loop. So a topologist's knot is a knotted circle, rather than a knotted curve with ends. For instance, the two knots in Figure 45 then look like Figure 46.

We can now rephrase the question in the caption to Figure 45 in knot-theoretic terms. One of the two 'knots' below is actually unknotted: you can transform it, continuously, into a perfect circle, with no crossings. The other is a genuine knot, and can't be untied without cutting the string. So the question is, which is the knot, and which the unknot? In fact, the more complicated-looking knot, on the right, is the one that unties. The other is a reef knot, and generations of boy scouts know that it does not untie. But that's not a mathematical proof, and what boy scouts mean by 'untie' includes things like free ends and the possibility of the knot

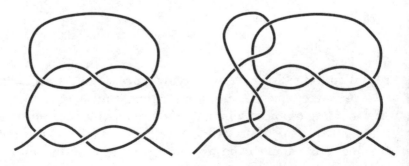

Fig 45 Pull on the ends – which one unknots?

Fig 46 Glue the ends to stop the knot escaping.

slipping. So topologists have to be more careful, and find solid logical proofs for things that seem obvious.

Knot theory has on occasion been ridiculed as pseudo-intellectual trivia. That's an understandable attitude if you don't know any mathematics and base your opinion on the everyday meaning of the word 'knot', but when it comes to mathematical concepts that's a rather silly way to think. It's like expecting quantum field theory to be solely about sheep. The mathematics of knots has turned out to be deep and difficult, and it has been a prime mover in the development of topology.

. . . .

Knot theory is useful in biology because DNA ties itself in knots. The knots are a relic of the twisted topology of the double helix. If you cut a length of DNA and join its ends together, two things can happen. Either you have joined each separate helix to itself, in which case you end up with two closed loops of single-strand DNA. Usually, these are linked to each other – impossible to separate without cutting. Or, you must have joined each strand to the other one. Now they form a single closed loop, and typically it is knotted.

If you can understand these knots and links, you can work out features of the biological process that did the cutting. And this is an important idea, because nature has to cut and rejoin DNA on a routine basis. The complex topology of the double helix forces this. Copying a DNA strand requires cutting it, disentangling it from its partner, building the new copy, then putting the cut strand back where it came from and rejoining it (see Figure 17 on p. 98). These processes are very complicated on a molecular level, but they make life possible. And since they are on a molecular level, it's not easy to observe them as they happen. Instead, they have to be inferred.

The method used to derive the double-helix structure of DNA is of little use here. It involves making a DNA crystal, illuminating it with X-rays and observing the resulting diffraction patterns. But cellular DNA is not a crystal. It is a free-moving molecule dissolved in a liquid. A new approach is needed to understand what DNA does in a cell.

The cellular machinery is not just the puzzle here: it also provides part of the answer. DNA is cut into pieces by special proteins, enzymes, known as topoisomerases. One way to find out what happens is to image the resulting strands using a very powerful electron microscope. Making these images required a new technique: coating the DNA strands with a special protein to make them thicker. The topology of those strands tells you useful things about the action of the topoisomerases. This in turn tells you things about the DNA. So you can get information about how topoisomerases do their work, and the effect they have on the DNA, by letting them cut up some DNA and seeing what shapes you get.

Biologists have discovered an effective way to keep this kind of investigation under control: perform the cutting operation on a closed loop of DNA, which can be constructed using standard techniques of genetic engineering and equipped with special regions whose code sequence can be recognised, and operated on, by a suitable enzyme. The result is either a DNA knot or two DNA loops that are linked together. The way the separate strands overlap can be observed with an electron microscope. Now you have a problem: you have a picture of a knot (similar to Figure 47), or a link – but which one is it? It may well be twisted and twirled in a way that makes the answer far from obvious. But topology can come to your rescue.

Knots are familiar, and seem simple, which suggests that they should be easy to understand. However, a quick glance at a book like the famous *Ashley Book of Knots* reveals the existence of thousands of different knots, and those are just the ones that turned out to be useful in the days of sailing ships, or are decorative, or can be used for party tricks. Distinguishing knots,

Fig 47   Twisted strands of DNA forming a Whitehead link.

and working out what happens to them when you make various types of change, are basic issues in topology. And they're *hard*.

For instance, although experiment made it clear thousands of years ago that a knotted length of string is different from an unknotted one, a solid logical proof of the existence of knots had to wait until the 1920s. The first big success of topological knot theory was a rigorous proof that the standard overhand knot, embedded in such a manner in a closed loop of string, cannot be untied. That is, no continuous transformation will convert it into an ordinary circular loop.

Why is this problem so difficult? It requires a proof that no transformation, however complicated and cunning, can do the job. It is much easier to analyse some specific transformation and see what it does, but this question can't be answered that way. In principle, you have to contemplate all possible transformations, and show that none of them unties the knot. In practice this is impossible, but there is a clever way to achieve the same result without considering infinitely many, ever more complicated transformations. The idea is to look for an invariant – a specific quantity or structure associated with any knot that stays the same when the knot is transformed. The invariant must also be something we can calculate, otherwise it will be pretty but useless.

Here's how an invariant does its job. Suppose, for example, that some cunning topologist invents an invariant, and when we calculate it, the answer is 3 for the overhand knot but 0 for the 'unknot', an unknotted loop. Then we can prove, with complete logical rigour, that no amount of twisting and turning and bending and stretching can convert an overhand knot into an unknot. Why? Because such transformations will always produce a knot whose invariant is 3. Since the unknot is not one of those, it can never be produced.

This is a straightforward and widely used idea, but there is a sting in the tail. First, the thing we're thinking of does actually have to be an invariant, and it may not be easy to find such a beast – or to prove its invariance. Second, it must be something we can calculate for the knot we start from, and for the one we hope to finish with; here, the unknot. Third, it must be different for the initial knot and the final one.

Despite these obstacles, topologists managed to invent some decent knot invariants, and they used them to solve basic problems. Prove the overhand knot is genuinely knotted. Prove that you can't transform an overhand knot into its mirror image. Prove that a reef knot is different from a granny knot, and both differ from the overhand knot. And so on.

Many of the more recent knot invariants are algebraic formulas, often called polynomials. There is a classic invariant of this kind, named for its inventor, James Waddell Alexander. You can calculate the Alexander polynomial from a picture of the knot. It is $x^{-1}-1+x$ for the overhand knot, and $-x^{-1}+3-x$ for the figure-of-eight knot. Since these are visibly different, so are the knots.

The Alexander polynomial solves some problems, but not others. It fails to distinguish reef from granny, and it can't tell the difference between an overhand knot and its mirror image. So topologists sought improved invariants. Ideally, what we want is an invariant that completely distinguishes all topologically different knots: if the knots differ topologically, then their invariants have to be different too. But that proved elusive. However, Vaughan Jones invented a pretty good new invariant in 1983. The Jones polynomial can distinguish reef from granny, and tell the difference between an overhand knot and its mirror image. Other mathematicians generalised his ideas, and we now have several powerful knot invariants.

Invariants are not the only way to approach questions about knots. An earlier idea was to construct knots from simpler building-blocks, and try to understand how the blocks worked and how they fitted together. This technique is useful in understanding the action of enzymes. John Horton Conway, a British mathematician famous for his unorthodoxy, imagination and playful approach to much of mathematics, invented one such structure, known as a *tangle*. By this term, he meant a piece of a knot whose ends are attached to a surrounding box. You can transform the bits of knot in any continuous manner, as long as they remain inside the box, but you have to leave their ends attached to the surface. The basic type of tangle consists of two separate strands, each with two ends, so there are four ends on the box, and the strands themselves are knotted and linked together in the interior of the box. In Figure 48 the boxes are shown as squares, but they are really three-dimensional.

There is a trivial tangle, in which the two strands are parallel and do not link or twist. Tangles can be combined by 'adding' two of them together. To do this, sit them side by side, join adjacent ends, and replace the two surrounding boxes by a single, larger box. This is the sense in which tangles are building-blocks, and by repeated additions, finally joining corresponding free ends to close the loop, you can make knots from them. The trivial tangle acts like zero: adding it to another tangle just extends one pair of strands, and has no topological effect.

In 1985 Nicholas Cozzarelli, a molecular and cell biologist at the University of California at Berkeley, and colleagues, applied tangles to a problem about DNA called site-specific recombination.[1] A site, in this context, is a short, two-stranded segment of DNA whose base sequence can be recognised by an enzyme. Two such sites can

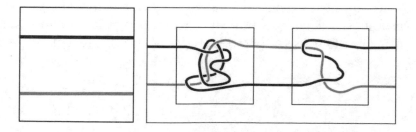

Fig 48   *Left:* trivial tangle. *Right:* Two tangles (inside small squares) and how to add them (outer rectangle). The shading distinguishes separate strands and has no other significance.

be recombined: the enzyme snips out corresponding segments of the two DNA chains that lie between them, swaps their ends round in some manner, and rejoins everything. Except for the business of surrounding boxes, this type of recombination has the same geometric effect as adding a new tangle to what was previously the trivial tangle.

Thus armed, we can attack a genuine biological problem: the action of an enzyme with the impressive name Tn3 resolvase. This is a site-specific recombinase. A recombinase is an enzyme that breaks DNA strands and joins them together again in a different way; it is site-specific if the place where it makes these changes corresponds to a specific short DNA sequence. Tn3 resolvase can undergo several successive reactions of this kind with double-stranded DNA, and the problem is to find out exactly how the strands are broken and rejoined. The topological clues, which at first sight are rather puzzling, make it possible to reconstruct how the enzyme acts.

Begin with a circular loop of two-stranded DNA, which topologically is the unknot. Allow the enzyme to act on it once, and the result is the so-called Hopf link. About once in 20 times there is a second reaction, leading to the figure-of-eight knot. Much rarer, but possible, is a triple reaction, which gives the so-called Whitehead link (see Figure 49).

The single-step reaction is illustrated in Figure 50. The unknotted loop of DNA is twisted up so that it divides naturally into three regions, each of which constitutes a tangle. These regions are the parallel strands inside the enzyme (shaded circle), the double-twist to its left and the rest of the DNA. The one that most concerns us here is the first of these, which is where the enzyme rearranges the DNA strands, in a manner that is shown schematically in Figure 50 as a crossing-over of the two strands. But that's just one guess at how the enzyme acts, and there are many other possibilities. The problem is to replace this schematic picture

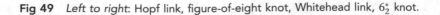

Fig 49 *Left to right:* Hopf link, figure-of-eight knot, Whitehead link, 6_2^* knot.

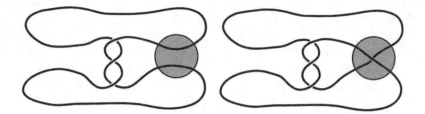

Fig 50 Action of enzyme Tn3 resolvase (grey circle) on a loop of DNA. *Left*: Before one application. *Right*: After.

by the one that actually occurs in nature, just by looking at the range of knots and links produced by the chemical reaction.

Cozzarelli's approach to this problem assumes that the enzyme makes the same change to the tangle topology each time it acts. This change can be viewed as tangle addition, with the same tangle being added each time. So we start with the unknot U and successively add a tangle X that represents the action of the enzyme, obtaining three tangle equations:

$U + X = $ Hopf link

$U + X + X = $ Figure-of-eight knot

$U + X + X + X = $ Whitehead link

We then solve these for X. This, remarkably, is possible, and there is a unique answer. The tangle X has to act in exactly the way the schematic figure assumes. But this is no longer just the simplest guess, but a consequence of specific biological assumptions about the enzyme action, verified by experiment.

Can we make a prediction, a new experiment that will test whether the theory is correct? We can. Even more rarely, there ought to be a fourth action of the enzyme, leading to $U+X+X+X+X$. Since X is now known, we can work out which knot or link this is. It turns out to be a knot with no familiar name, which topologists call 6_2^* (see Figure 49). The prediction, then, is that even less frequently than the triple reaction, we will observe the knot 6_2^*. And observations do indeed detect precisely this knot, with roughly the predicted degree of rarity.

• • • • • • • • • • • •

Mathematical issues that sound fairly similar to those that arise with knots, but technically are very different, arise in a related area of molecular biology: protein folding. Although a protein is conceptually a chain of amino acids, in reality the chain folds up in a complex way under the influence of molecular forces. The basic point here is that the DNA, interpreted as amino acids, does not 'contain the information' that tells the protein how it should fold. Instead, the protein folds automatically, in response to the chemicals in the surrounding medium, the activity of special molecules called chaperonins which nudge it into particular configurations, temperature, and other factors.

Biologists often consider that anything passively obeying the laws of physics and chemistry is merely part of the background against which biology works. From that point of view, all that matters is that the physics does whatever the physics does. It can be taken for granted, even ignored. An elephant pushed off a cliff will fall – but that's gravity, not biology. However, it is not possible to assign protein folding to this kind of background operation of physical law, because in principle proteins can fold in a huge variety of ways. Even if in practice they fold in only one way, we still need to know which shape arises, because a protein molecule's shape is one of the main features that determine its biological function (or functions). Think of haemoglobin, which acts rather like molecular tongs, picking up a molecule of oxygen and putting it down again. If it weren't the right shape, it couldn't do the job. Figure 51 shows how haemoglobin folds, and its two slightly different configurations.[2]

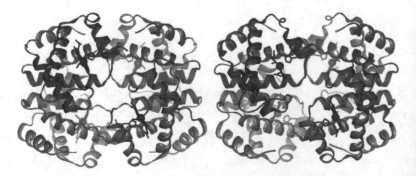

Fig 51 The two shapes of haemoglobin. *Left*: oxygen binds to regions shown as small dark spheres. *Right*: oxygen is released.

By the way, I'm not suggesting that haemoglobin is the *only* molecule that can transport oxygen.[3] Tongs can have differently shaped handles without it impairing their function. Similarly, many other proteins could in principle transport oxygen. But any suitable molecule has to have a shape that lets it behave like oxygen-tongs. I mention this because it is sometimes suggested that haemoglobin is too complicated to have evolved – as if this *specific* molecule were a target for evolution to aim at. On the contrary: evolution is opportunistic, and will settle for anything that works.

This role of shape is not just of theoretical importance. Many diseases, among them Creutzfeldt–Jakob disease, mad cow disease (BSE) and probably Alzheimer's disease, may be caused by misfolded proteins. The ability to deduce the shape of a protein from its amino acid sequence would be a huge step forwards in biology, because sequencing DNA is now cheap and easy (if you have the rather expensive equipment and skills required) but working out the shape of a complicated protein is very hard.

The process was a huge puzzle until around 1990, when Joseph Bryngelson and Peter Wolynes at the University of Illinois devised a mathematical formulation in terms of an 'energy landscape'. The forces that act between atoms and electrons in a molecule imply that any configuration of an amino acid chain (or any other molecule, for that matter) has a definite amount of energy. The mathematics and physics of dynamical systems imply that the molecule will behave in a manner that tries to make its energy as small as possible. Take an elastic band, and drop it on the desk. It tends to take up one particular unstretched shape. However, you can stretch the band into all sorts of shapes by pulling on different bits of it with your fingers. As you stretch it, though, you feel a certain amount of resistance. The more you stretch it, the harder you have to pull. What's happening here is that in order to stretch the band, you have to increase its elastic energy. All the stretched shapes have more energy than the natural unstretched shape, and you have to do work to provide the extra energy. So the unstretched shape is the shape with the least energy.

It's much the same with a protein molecule, but the different amino acids make it more like an elastic band with all sorts of lumps and bumps. Nevertheless, the stretched-out-straight shape has a lot of energy, and the protein molecule prefers to contract into a less energetic shape. In this way, nature can produce a specific shape of molecule by stitching amino acids together in turn, following the genetic instructions, and then just allowing the chain to fold itself into the required shape. Minimisation of energy does all the hard work, and the genes don't need to tell the protein how to fold.

In the metaphor of an energy landscape, variations in energy of the conceivable configurations can be seen as hills and valleys, creating a conceptual landscape in which height corresponds to energy, and location in the landscape corresponds to the configuration concerned.

It is not practical to determine the actual shape of the protein by considering all possible shapes, working out the energy for each and seeing which is smallest, because the range of possible shapes is absolutely gigantic. It would be like trying to predict the shape of an elastic band by considering every conceivable shape, however implausible, computing the energy, and seeing which is smallest. This is not the only difficulty. I said that the molecule always 'tries to make its energy as small as possible', but that is an oversimplification. I should have said 'as small as possible, *compared with any nearby configuration*'. The energy landscape may have a local depression, even though the lowest point on the entire landscape is far below that.

It is not only mathematicians who have trouble here. The molecule itself is much like the elastic band. It has no idea where the lowest point in the landscape is, so it heads downhill and finds out where that leads. If it leads into a local depression, the molecule gets trapped in the wrong configuration and the protein can't do its job. In the late 1960s, Cyrus Levinthal, then a molecular biologist at Columbia University, realised that the energy landscape for a typical amino acid chain can have a gigantic number of possible local depressions.[4] Suppose that the chain consists of 300 amino acids – which if anything is on the small side – and that the chemical bond linking each to the next can adopt one of a mere three stable angles. These angles are more or less independent, so the number of possible combinations is 3^{300}, roughly 10^{143}. This leads to Levinthal paradox: as a real protein chain folds, it cannot reach the 'correct' configuration by trying all possible configurations in turn. The universe will not last long enough.

The deduction is not that protein chains perform miracles, but

that they don't do it that way. A popular theory holds that
evolution has eased the path: to ensure correct folding, the chains
that are found in biologically significant proteins are not typical.
Instead of having rough energy landscapes, with lots of bumps and
hillocks, their landscapes form steep-sided funnels, with an obvious
route down into the depths where the desired configuration lurks
(see Figure 52). More precisely, they are made from a series of such
funnels, each feeding the configuration into the next funnel along
a specific path.[5] The funnels are linked by saddle points – places
which are local energy peaks in some directions and local
depressions in others, like a mountain pass in a conventional
landscape. The pass is the lowest point through which a traveller
can cross a mountain ridge, so the path goes up to the pass and
then descends, but the ridge itself rises upwards from the pass.

The idea is that natural selection will favour proteins with
simple energy landscapes like this. Not because evolution somehow
knows what the landscape looks like, but because molecules that
often fold into a shape that doesn't work will make the host
organism less likely to survive, so these proteins will be weeded out.

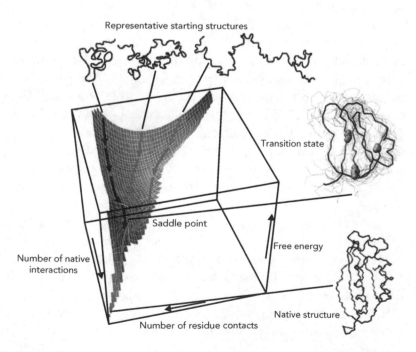

Fig 52 Energy landscape of a highly simplified model of a small protein, showing
a deep funnel.

Even if this theory were right, the mathematics of protein folding would not become simple and straightforward, for other reasons. But at least we would understand how the molecule gets itself folded correctly.

· · · · · · · · · · · ·

Many different software packages make some kind of a stab at predicting how a protein will fold, using a mixture of mathematical principles and informed guesswork. A popular one is Rosetta, which harnesses the power of idle computers worldwide through its Rosetta@home project. This uses the Berkeley Open Infrastructure for Network Computing (BOINC) to carry out huge computations on a network provided by more than 80,000 volunteers, who allow the use of their home computers when they are not otherwise occupied. But in 2008 Seth Cooper and colleagues went one better, by turning protein folding into a multiplayer online computer game: Foldit. Players compete with one another and progress through increasing levels of difficulty, looking for the right way to fold a given protein.

Doing science using a computer game might seem an absurd piece of deference to popular culture, but it makes very effective use of something that humans have in abundance and computers lack: intuition. The human brain is very good at spotting patterns, even ones that it doesn't consciously recognise. In 2010 Cooper's team reported that 'top-ranked Foldit players excel at solving challenging structural refinement problems'.[6] The element of collaboration seems to add further power to the human brain's intuitive understanding of three-dimensional shapes, and the competitive instinct provides motivation – as it does even in conventional science.

Foldit benefited from advice from professional game designers. Players progress by solving a series of puzzles, initially based on proteins whose structure is known, but not publicly available. Along the way they learn the many technical terms and techniques based on Rosetta, such as 'combinatorial side-chain rotamer packing', but using more friendly terminology – in this case, 'shake'. Then they can progress to unsolved structures, with potential contributions to serious science.

Foldit is an intriguing example of a growing trend in science:

getting the general public to participate in scientific research using distributed computing over the Internet. The problems have to be set up in an easily accessible way, but once this is done, a vast amount of computational power *and* human input becomes available – at very little cost. It's a trend that could go a long way. As Zoran Popović, a project member, has said: 'Our ultimate goal is to have ordinary people play the game and eventually be candidates for winning the Nobel prize.'

Foldit certainly beats *Grand Theft Auto* as a constructive way to pass the time, though it might not give you quite the same adrenaline rush.[7]

13 Spots and Stripes

Painters, poets and writers have long been captivated by the extraordinary beauty of animals in the wild. Who could fail to be moved by the power and elegance of a Siberian tiger, the ponderous enormity of an elephant, the haughty pose of a giraffe or the pop-art stripes of a zebra?

Yet each of these animals began life as a single cell, a fusion of sperm and egg. How do you cram an elephant into a cell?

When the paradigm of DNA as information was at its height, the answer seemed simple. You don't. What you cram into an egg is the *information* required to make an elephant. Since that is in molecular form, an awful lot of information can be confined to the interior of a single cell.

If you do the sums, however, it's clearly not that straightforward. An elephant has many more cells in its body than its DNA has bases. The cells come in many different types. They have to be fitted together in the right way ... Have you ever considered how complicated an accurate, cell-by-cell map of an elephant would be? Let alone putting in the complex organelles and cytoskeleton inside each cell.

Part of the answer is the background of physical and chemical laws (see p. 192). These laws operate automatically; in fact, you can't stop them doing so. A tiger's DNA doesn't have to contain information about how chemical bonds fit together, how cells that are sticky adhere to one another, how electrical impulses pass along its neurons. All these things are implicit in the laws of nature.

However, that's only part of the answer. It's not enough to assign anything you don't find in DNA to the action of physical laws, because this fails to answer any of the big questions. In particular: how does DNA regulate the complex processes of chemistry, to turn that egg – seething as it may be with potential – into a gigantic striped cat?

This is where mathematics can play its part in the great scientific enterprise. We don't, yet, have accurate mathematical models for all the processes that convert egg into tiger, but we do have models that provide insight into various stages of that process. Mathematics can help us understand simple features of a growing embryo, such as gastrulation, where a spherical mass of tiny cells turns inwards on itself, the first stage in giving the eventual animal an inside and an outside. There are many other applications of mathematics to biological development, but I want to skip a few stages to look at the most obvious features of many animals: their markings.

· · · · · · · · · · · · ·

The mathematician who opened up this area was an Englishman, Alan Turing. Turing is famous for his wartime work at Bletchley Park breaking the Enigma code, for developing the Turing test for artificial intelligence, and for establishing the undecidability of the halting problem for Turing machines. (That is, there is no systematic way to determine whether any given computer program will terminate with an answer, or go on for ever – for example, by repeating the same instructions over and over again indefinitely.) From these activities it might appear that Turing was a pioneer in computer science and cryptography, which is true, and that he specialised in these areas – which is false. Another mathematical interest of his was the markings on animals. Spots, stripes, dappled patterns . . .

For half a century, mathematical biologists have built on Turing's ideas. His specific model, and the biological theory of pattern-formation that motivated it, turns out to be too simple to explain many details of animal markings, but it captures many important features in a simple context, and points the way to models that are biologically realistic (Figure 53, see over).

It all started in the early 1950s, when Turing became puzzled about the geometry of animal form and markings: the stripes on tigers and zebras, the spots on leopards, the dappled patches on

Fig 53 *Left*: Boxfish. *Right*: Calculated Turing pattern.

Friesian cows. Although these patterns do not display the exact regularity that people often expect from mathematics, they have a distinct mathematical 'feel'. We now know that the mathematics of pattern formation can produce irregular patterns as well as regular ones, so the irregularities are not evidence that mathematical models of animal patterns are wrong.

Turing presented his theory of pattern formation in a celebrated paper entitled 'The chemical basis of morphogenesis', published in 1952.[1] He modelled the formation of animal markings as a process that laid down a 'pre-pattern' in the developing embryo. As the embryo grew, this pre-pattern became expressed as a pattern of protein pigments. He therefore concentrated on modelling the pre-pattern. His model has two main ingredients: reaction and diffusion. Turing imagined some system of chemicals, which he called morphogens, 'form-generators'. At any given point on the part of the embryo that eventually becomes the skin – in effect, the embryo's surface – these morphogens react together to create other chemical molecules. The chemicals and their reaction products can also diffuse, moving across the skin in any direction.

Chemical reactions require nonlinear equations, ones in which – for example – twice the input does not yield twice the output. Diffusion can sensibly be modelled by simpler linear equations: twice as much of some molecule, diffusing from some given location, leads to twice as much everywhere. The most important result to emerge from Turing's 'reaction–diffusion' equations is that local nonlinearity plus global diffusion creates striking and often complex patterns (see Figure 54). Many similar equations can produce patterns, not just the specific ones proposed by Turing. What distinguishes them is which patterns occur in which circumstances.

Fig 54 *Above*: Regular Turing patterns: spots and stripes. *Below*: Less regular Turing patterns.

. . . .

Hans Meinhardt, at the Max Planck Institute for Developmental Biology in Tübingen, has made extensive studies of many variants of Turing's equations. In his elegant book *The Algorithmic Beauty of Seashells* he examines many different kinds of chemical mechanism, showing that particular types of reaction lead to particular kinds of pattern. For example, some of the chemicals inhibit the production of others, some activate the production of others. Combinations of inhibitors and activators can cause chemical oscillations, resulting in regular patterns of stripes or spots. Meinhardt's theoretical patterns match those found on real shells.

Use of the word 'pattern' does not imply regularity. Many striking patterns on seashells are complex and irregular. Some cone shells have what seem to be random collections of triangles of various sizes (Figure 55, see over). Mathematically, patterns of this kind can arise from Turing-like equations; they are fractals, a complex kind of geometric structure popularised by the French-American mathematician Benoît Mandelbrot working at Yale in the 1960s. Fractals are closely associated with dynamical chaos – irregular behaviour in a deterministic mathematical system. So the cone shell combines mathematical features of order and chaos in one pattern.

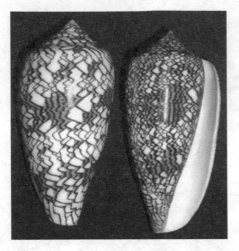

Fig 55 Cone shell pattern.

The Scottish mathematician James Murray of the Universities of Washington and Oxford has applied Turing's ideas, suitably modified and extended, to the markings on big cats, giraffes, zebras and related animals.[2] Here the two classic patterns are stripes (tiger, zebra) and spots (cheetah, leopard). Both patterns are created by wave-like structures in the chemistry. Long, parallel waves produce stripes. A second system of waves, at an angle to the first, can cause the stripes to break up into series of spots. Mathematically, stripes turn into spots when the pattern of parallel waves becomes unstable. Pursuing this led Murray to an interesting theorem: a spotted animal can have a striped tail, but a striped animal cannot have a spotted tail. The smaller diameter of the tail leaves less room for stripes to become unstable, whereas this instability is more likely on the larger-diameter body.

In 1995 the Japanese scientists Shigeru Kondo (Kyoto University Centre for Molecular Biology and Genetics) and Rihito Asai (Kyoto University Seto Marine Biological Laboratory) used Turing's equations to make a startling discovery about the colourful tropical angelfish *Pomacanthus imperator*. Along two-thirds of its body run parallel stripes of yellow and purple. Stripes are an archetypal Turing pattern, but there is an apparent technical difficulty. In this particular case the mathematics of Turing patterns makes a surprising prediction: the stripes of the angelfish have to *move*. Stable steady patterns are not consistent with the mathematical formalism.

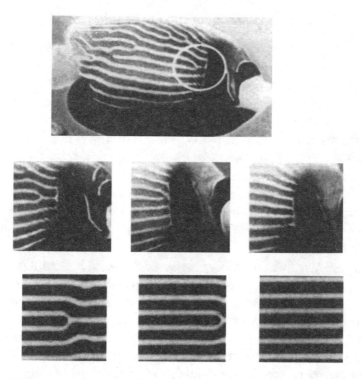

Fig 56 Moving stripes on angelfish. *Top*: Angelfish whose stripes have a Y-shaped branch (circled). *Middle*: Observations at intervals of 6 weeks show that the branch moves. *Bottom*: Theoretical movement according to Turing's model.

Moving stripes? It seemed bizarre. However, Kondo and Asai decided to keep an open mind. Maybe angelfish stripes *do* move. To find out, they photographed specimens of the angelfish over periods of several months. They found that the stripes do, in fact, migrate across its surface (see Figure 56).[3] As they move, certain defects in the pattern of otherwise regular stripes, known as dislocations, break up and reform exactly as Turing's equations predict.

. . . .

Pattern formation is predicted by a variety of mathematical models, many of which give rise to the same catalogue of possible patterns – those that occur in nature as stripes in ocean waves, on tigers and on angelfish, for instance. There ought to be some deeper, general reason for these similarities – indeed, for the patterns themselves. There is, and it's called symmetry breaking.

In normal parlance, something is symmetric if it has an elegant, balanced form. More specifically, we talk of an object having bilateral or mirror symmetry if it looks the same as its reflection in a mirror. The human body comes close to this, although a person and their mirror image differ in minor details, and the arrangement of internal organs can differ markedly.

There are more complicated types of symmetry; my favourite example is the starfish. The most common starfish found around the British Isles has five arms, each arm more or less identical to the others, all of them arranged in something close to a five-pointed star. A starfish does have mirror symmetry, but the most obvious symmetry is not a reflection, but a rotation: each arm is separated from the next by 72°, one-fifth of a turn.

Symmetry is not about parts of some shape being similar to one another: it is about the effect of some transformation on the whole shape. Does the shape appear to stay the same when reflected? If so, it has mirror symmetry. Does it appear to stay the same when rotated? Then it has rotational symmetry. As we've seen, this idea has become a formal mathematical theory: group theory, so named because any two symmetries combine to yield another one. The set of all symmetries of an object is its symmetry group.

Not only can the solutions of mathematical equations have symmetries; the equations *themselves* can have symmetries. The algebraic formula $x+y$ is symmetric in the two numbers x and y: if you swap them, you get the same sum. More generally, an equation is symmetric if applying some transformation to the symbols yields the same equation. By the middle of the last century's first decade, Albert Einstein had already noted the importance of this type of symmetry in the laws of physics, the basic equations for space, time, energy and matter. He insisted that these laws must be the same at every point of space and at every instant of time.

It may seem surprising that solutions can have less symmetry than the equations they solve. Working in Paris a few years before Einstein, Pierre Curie suggested that this should not be possible because of a simple physical principle: symmetric causes produce equally symmetric effects. Curie's principle seems to rule out a change of symmetry from equations ('causes') to solutions ('effects'). But if the universe itself had that much symmetry, looking exactly the same in every location and at every instant, it would be uniform and unchanging, like a universe composed entirely of cosmic custard. So there must be some escape clause. There is: it's symmetry breaking, and its consequences are both beautiful and far-reaching.

How can symmetry break? More precisely, how can a solution of an equation have less symmetry than the equation itself? If an equation is symmetric, then some transformation of the symbols yields the same equation. So applying that transformation to a solution also yields a solution.

However, it need not be the *same* solution. This is the loophole that makes symmetry breaking possible.

In our algebraic example, the sum $x+y$ of two numbers is symmetric in x and y. If we swap x and y, it becomes $y+x$; formally, this looks different, but it always yields the same answer. That statement remains true for specific choices of x and y, such as 17 and 36. With these choices $17+36=36+17$. But this does not imply that 17 and 36 remain the same if we swap them. On the contrary, 17 turns into 36, which is different. Of course, 36 correspondingly turns into 17 – but again, those two numbers are different. So the solution $x=36$, $y=17$ is different from the solution $x=17$, $y=36$. Curie's principle is correct as it stands for equations that have only one solution. But when there are many solutions, which turns out to be very common, the principle as originally stated fails. Instead, all we can assert is that whenever we have a solution, we can obtain other solutions by applying symmetry transformations.

. . . .

We've seen that symmetry is very common in mathematical models of the natural world, so symmetry breaking should also be common. And so it is. In fact, it provides a very general mechanism for the formation of nature's patterns. Those patterns are the explicit realisations, in specific physical systems, of the abstract symmetries that are implicit in the laws that describe those systems.

Multiple solutions open the door to symmetry breaking. What shoves the mathematics through that door is instability.

On the left of Figure 57 (see over) is a satellite photo of part of the Rub' al-Khali desert in Saudi Arabia, often known by its English name, the Empty Quarter. The stripes are huge sand dunes. Although there are irregularities, the stripes are pretty much parallel to one another and equally spaced. The pattern is caused by strong

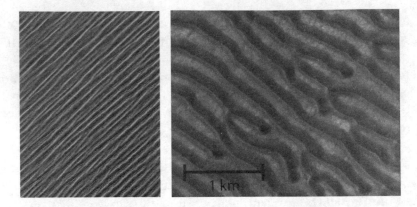

Fig 57 *Left*: Longitudinal dunes in the Empty Quarter (width is roughly 40 km).
Right: Transverse dunes on Mars.

and steady trade winds, which in this case blow in the same
direction as the stripes: accordingly, these are known as
longitudinal dunes. Striped dunes can also occur when the wind
blows at right angles to the stripes: the result looks much the same,
but the different mode of formation requires a different name:
transverse dunes. The right-hand photo shows transverse dunes on
Mars.

There are many other dune patterns, including wonderful
crescent shapes and stars, but striped dunes are the simplest. Now,
it might seem that the strong patterns in the dunes must reflect
equally strong patterns in the way the wind blows, but longitudinal
and transverse dunes typically form when the wind is steady. In
fact, the steadier the wind, the more regular the dune pattern.

Of course once the dunes have formed, they affect the wind in
their locality, but you don't need a striped wind to create striped
dunes.

Why not?

To avoid confusion, let me focus on transverse dunes.
Longitudinal dunes have a similar, perhaps simpler, explanation,
but the transverse ones will serve my purpose better later on.

Imagine a perfectly flat desert, over which a steady wind blows,
at the same speed and in the same direction everywhere. No such
desert can exist in reality, of course, but this ideal case homes in on
the essence of the puzzle: how a uniform wind can lead to a striped
desert. The key is symmetry, and how it breaks. My idealised desert
is very symmetric. In fact, the only departure from the symmetry of

a mathematical plane is the existence of a preferred direction, that of the wind. So the transformations that preserve the system include no rotations, and the only reflectional symmetries occur for mirrors aligned with the wind direction. However, I can slide the entire desert north, south, east or west, and the system – and therefore also its mathematical representation – will look exactly the same.

If the behaviour of the sand in response to the wind were as symmetric as the system itself, there would be no patterns. The state of the sand, and in particular the height of the desert surface, would be identical at every point. So the sand would stay flat, and the symmetry of the system would not break.

If we inject just one element of realism, however, this picture changes dramatically. Sand is not smooth: it comes in tiny grains. Those grains poke above the surface in places, and there are gaps between them. The surface departs from perfect planarity by a very tiny amount. And those departures are pretty much random. Such a system has no symmetry at all. However you transform the desert, the sand grains will not repeat the exact same pattern.

The difference is tiny, but what actually happens in a (fairly) flat desert subjected to (fairly) constant winds is huge and dramatic. Great mounds of sand appear, many thousands of times the size of the sand grains that cause the departure from exact symmetry. And, very commonly, the dunes have large-scale patterns: transverse dunes, for example, are arranged in regularly spaced parallel stripes like waves on a beach. Ocean waves move, and so do transverse dunes: they move at right angles to the stripes. But they do so very slowly, as sand is blown off the crest of each dune and is deposited somewhere ahead of the crest.

Parallel rows of dunes have quite a lot of symmetry. They form a striped pattern in the plane, and this can be slid sideways along the direction of the stripes. It is also unchanged if it is slid perpendicular to the stripes, through a distance that is any integer multiple of the distance between adjacent stripes – the actual spacing, twice that, three times, and so on.

This is remarkable. The symmetry of the typical pattern of behaviour resembles neither that of the perfect idealised model, with complete translational symmetry, nor the small but total asymmetry of real sand grains. Instead, it lies somewhere in between. And it turns out that the idealised model can reproduce exactly that kind of pattern, if it is tweaked very slightly to mimic the random *but tiny* deviations from perfection introduced by the granular nature of sand. In such a model, the sand remains very close to a perfect plane provided the wind speed is sufficiently low – low enough not to disturb any sand grains. But for higher wind speeds this undisturbed state becomes unstable. Any tiny imperfection, however small, grows. If a grain of sand pokes up slightly more than its neighbours, the wind picks it up and blows it somewhere else. The resulting hole creates a bigger difference in height, and the grains on either side become more exposed and also get blown away. The hole grows, and the displaced sand piles up.

However, the sand does not pile up at random. Instead, feedback between the shape of the desert surface and the movement of the wind homes in on a stable pattern: waves of sand and waves of wind. In the right range of wind speeds, that pattern is transverse dunes.

Where has the symmetry of the system *gone*?

That could be a silly question. Symmetry isn't a physical substance that can't be annihilated without creating something else. Symmetry is a concept, a property. But in this case it's not a silly question, because the missing symmetry *has* gone somewhere. It exists as unrealised potential. That pattern of stripes could have formed in any position. Its actual position was a consequence of the first grain of sand starting to move. Potentially, that grain of sand might have been anywhere in the desert. If so, the entire process would have taken place some distance away – so the crests of the potential dunes would have formed somewhere in the troughs of the actual ones. The symmetry of the system is not so much broken as *shared* among the entire set of possible solutions.

A model that predicts the formation of transverse dunes cannot specify the exact location of their crests and troughs. If we moved the entire pattern forward ten metres, it would also satisfy the equations of the model. In fact, if we waited long enough the wind would actually move the pattern into that position. At any given moment, the location of the crests and troughs depends on the past history of the dunes.

There are profound mathematical techniques for calculating stabilities and working out which patterns appear when fully symmetric states become unstable. They are very technical, but in general terms they suggest that nature prefers not to break symmetry much more than it has to. Typical patterns arising in a symmetric system through spontaneous symmetry breaking tend to have a lot of symmetry rather than just a little. This statement can be made precise, in any particular instance: if taken too literally it is false, but on the whole not by much.[4] When it is false, the mathematics tells us what to expect instead.

• • • • • • • • • • • •

What about animal *form*? Form and pattern are two aspects of the same thing: morphology. Both form and pattern seem to be set up in embryos through a chemical pre-pattern induced by a morphogen. The pre-pattern sits there until the organism reaches an appropriate stage of development, at which point the varying chemical concentrations in the pre-pattern trigger either the formation of pigment proteins, creating visible patterns, or cellular changes, creating form.

There are disagreements about the precise mechanism that sets up the pre-patterns and about the precise mechanism that turns pre-patterns into visible patterns or form. Many of the chemical changes involved clearly have a genetic component – particular genes 'switch on' simultaneously in blocks of cells, stimulating the production of some pigment, or causing the cells to modify their mechanical or chemical properties. Pre-patterns alone cannot explain morphology: their interaction with genes presumably can.

Meinhardt has applied Turing equations, together with simple genetic 'switching' ideas, to the formation of somites in developing vertebrate embryos.[5] Somites are equally spaced pairs of blocks of differentiated tissue which eventually form the basis of the backbone. They come into being, one pair at a time, starting from the head end of the animal. On general mathematical grounds, however, Meinhardt was led to a counter-intuitive mathematical prediction. The diffusing chemical waves that trigger the formation of somites should originate not from the front of the animal, but from the back.

Why? Imagine ocean waves carrying floating debris up onto a beach, as the tide is going out. One wave reaches up to the top of the beach and deposits a strip of driftwood and seaweed. The next wave doesn't reach quite as far, thanks to the falling tide, so it leaves a line of debris further down the beach. Step by step, the waves travelling *up* the beach create a series of stripes of debris that accumulates *down* the beach. That way, the beach between the waves and the debris remains pristine, so the waves can deposit more debris using exactly the same process. The existing debris doesn't get in the way.

For somites, the waves are waves of concentration of some morphogen, and the debris that is deposited is a series of genetic 'switches' that change the state of the relevant cells. Again, the somites that have already formed would interfere with incoming waves unless the waves came from the back. Meinhardt made this prediction over fifteen years ago, as a consequence of Turing's mathematical scheme. The new-found ability to make genetic switching visible has shown that he was basically correct.

Despite these remarkable achievements, Turing's equations for animal markings have unsurprisingly – given the pioneering nature of his work – not been a complete success: they often fail to predict experimental details, such as what happens when you grow organisms at different temperatures. Turing was the first to attempt this kind of modelling, and he kept the model as simple as he dared: at that time solutions had to be calculated by hand. His theory has spun off many more sophisticated modern descendants, each of which attempts to address such deficiencies.

Whatever the details, though, the spatio-temporal patterns of activity of the genes are taken, virtually unchanged, from Turing's mathematical pattern-book. So it looks as if DNA guides morphogenesis along certain lines, but what then happens is heavily dependent on the laws of physics and chemistry too, hence upon *context*.

We know a lot about how DNA makes proteins, but in comparison we know very little about how those proteins are marshalled together to create an organism. The problem of biological development is one of the biggest scientific challenges we face. How do organisms regulate their own growth patterns? What defines an animal's body plan? How is its form transferred from the DNA drawing-board to the developmental assembly line? The answers will involve chemistry, biology, physics and mathematics. And they will be nowhere near as straightforward as just obeying a list of genetic instructions.

• • • • • • • • • • • •

The availability of powerful computers provides an alternative to classical continuum models like Turing's. Instead of approximating animal tissue by an infinitely divisible region of space, we can model the tissue cell by cell. We can study how cells affect their neighbours, how their internal dynamics and genetic regulatory systems conspire with the external world to determine their fate.

At the start of this chapter I mentioned gastrulation, the stage in embryonic development at which the growing mass of cells pretty much turns itself inside out. This process looks mathematical: it typically starts when a circular arc forms on the surface of a hollow sphere of cells; this becomes the lip of the inwardly folding portion of the surface. Many people have built mathematical models of gastrulation, but biologists know that the entire process is regulated by a few genes, and what they'd like to know is how the genetics interacts with the geometry.

In the 1990s, working at the Artificial Intelligence Laboratory in Zurich, Peter Eggenberger Hotz devised a series of mathematical models that incorporated the role of genes.[6] A typical strategy is to start with a grid of cells, represented as adjacent spheres, and write down a list of dynamical rules for how the genes present in these cells interact with those in neighbouring cells. A gene is activated (or inhibited) only when the total concentration of incoming signalling molecules exceeds a threshold level. The cell then responds to the activity of this gene, either by sending out its own signalling molecule, making a cell adhesion molecule that connects it to a nearby cell, or by equipping that cell with a receptor that can respond to an incoming signalling molecule. Additionally, the cell may respond by dividing or by dying.

The model is then simulated on a computer, following the dynamical rules and seeing what transpires. Depending on the choices made for the dynamics, the collection of cells may grow and develop in interesting ways. The shape that the mass of cells takes up is calculated using a fairly realistic set of equations for a collection of objects that interact through stickiness and elasticity, two key properties of a real cell.

As the mass of simulated cells develops, genetic signals cause the cells to change position in space. These changes in turn affect the activity of the genes and the signals they produce. The feedback loop between genes and form leads to the eventual shape of the mass of cells. The model can be used for many purposes, for example to explore the effect of a morphogen on morphology. One version of the model mimics gastrulation in a hollow sphere of cells.

Because the entire model exists in a computer, it is possible to investigate features of the structure that are difficult or impossible to observe in a real developing embryo, such as the concentrations of signalling molecules at various locations. This is a major advantage of *all* models. The corresponding disadvantage is that they are not the real system. As Eggenberger remarks, 'Putting evolutionary techniques on firm ground, where the mechanisms can be understood, is itself a major reason to investigate the potential of such systems.'

14 Lizard Games

· ·

A male lizard has secured himself a female, and soon the pair will
mate. She seems to like his sky-blue throat, and the two of them are
often seen walking out together. But suddenly this lizard equivalent
of marital bliss is shattered by an intruder. He is bigger and stronger
than the blue-throated lizard, and his throat is orange. He threatens
the blue-throated lizard, hoping to drive him away and steal his
female. When the blue-throated lizard resists, the orange-throated
one attacks.

This turns out to be a tactical error on both their parts, because
while they are engaged in battle, a smaller lizard with a yellow
throat sneaks up on the female and mates with her.

This reptilian soap opera is played out over and over again on
some of the many islands that dot the western coast of North
America. Since the males are competing for the same female, we
would expect them all to belong to the same species, and despite
their different colours, they do. The males are different 'morphs' –
types – of the common side-blotched lizard, scientific name *Uta
stansburiana*.

Their *Heat*-magazine-style mating strategies have remarkable
consequences.

Barry Sinervo, working from his lab at the Department of
Ecology and Evolutionary Biology at the University of California in
Santa Cruz, has been following the patterns of heredity in one
population of common side-blotched lizards since 1989. Because
they live on the same island and interbreed, they can be assigned to
a single species with some confidence. As we have just seen, the
males of this species occur in three distinct morphs, and the main

distinguishing feature is the colour of the lizard's throat, which can be orange, blue or yellow. The three morphs also differ in size, with the orange-throated ones tending to be larger and the yellow-throated ones smaller.

These colours seem to be an example of what Darwin called sexual selection: they relate not to general survival characters, but to female preferences when choosing mates. Anything that the ladies prefer tends to become more common, because it gets passed to their offspring. The peacock's gigantic, brightly coloured tail and the gaudy and bizarre decorations of birds of paradise are familiar examples.

Year in, year out, each morph follows its own particular mating strategy. Blue-throated male lizards form strong pair bonds with their females; orange-throated and yellow-throated ones don't. Orange morphs, which are the strongest, fight blue ones and take away their females. Yellow ones are coloured much like females, and thanks to this disguise they can take advantage of fights to approach without causing alarm, and mate with the disputed female. The blue ones rely mostly on strong pair bonding, tend to lose out to the stronger orange ones, but can defeat the yellow ones. In simplified terms:

- orange beats blue,
- blue beats yellow,
- yellow beats orange.

So orange is fitter than blue is fitter than yellow is fitter than ... orange. So much for 'survival of the fittest'.

What on earth is going on here? How does this evolutionary competition pan out, and how does it fit into a Darwinian picture?

. . . .

One of the biggest problems with the theory of evolution is that everyone thinks they understand it. But it would be much fairer to say that no one understands it – not even evolutionary biologists. Evolution is extremely complex and extremely subtle. It is not just a matter of the 'best' or 'fittest' creature winning the battle for survival. If it were, then one colour of male common side-blotched lizard would have displaced the other two long ago.

Is this a sign that, for these lizards, evolution doesn't work? It certainly looks that way if you take the slogan 'survival of the fittest' literally and interpret it naively. Even if these lizards' strange mating habits did evolve, they make it clear that the slogan is not a very good one. For this kind of reason, biologists avoid it.

Survival as such is not the main criterion for natural selection to operate: what counts is whether a creature manages to breed. It obviously has to survive to breeding age if it is going to have any chance of breeding, but it may fail to breed even if it has survived. The lizards demonstrate this point admirably. In innumerable species, only a few males breed, and they spend much of their time fighting the others to protect their own conjugal rights.

Moreover, the concept of 'fitness' in evolution is slippery. It's not just a matter of assigning a fixed measure of fitness to each creature, and comparing the numbers to determine which one will out-compete the other. If it were that simple, the planet would end up with exactly one species: the fittest one. But life on Earth is not like that. Natural selection is not like that either, and neither is biological fitness.

Evolutionary biologists have a love–hate relationship with fitness. They are aware of its shortcomings, but some of them feel that if these can be circumvented, the concept adds predictive value to evolutionary theory. What quickly becomes clear, if you follow that line of thought, is that the fitness of an organism does not depend just on the organism: it also depends on the context. In a game of golf, Tiger Woods would be fitter (much fitter) than me, but in a game of mathematics I would be fitter than Tiger. Paula Radcliffe would be fitter than either of us if we were running a marathon. What makes organisms 'fit' depends on which game is being played, as well as on the organisms playing it.

If we insist on understanding the behaviour of the three lizards in terms of some concept of fitness, tailored to their particular games, then the definition of fitness must depend on which game the creatures are playing. Yellow-throated males will lose out against orange-throated ones in a straight fight, but they can win if the orange-throated one is distracted by a battle. The blue-throated lizards' devotion to their mates will defeat the yellow-throated lizards, but not the orange. And the orange-throated lizards can beat the blue-throated ones in a fight, but they have difficulty keeping an eye on those sneaky yellow-throats.

• • • • • • • • • • • •

I referred to the lizard soap opera as a game – and so it is, in two senses. First, it has a lot in common with a game that children like to play. Second, both the lizards' game and the children's one can be modelled by a specific mathematical process, which happens to be called a game. Accordingly, the relevant area of mathematics is called game theory.

The childhood game I have in mind is scissors–paper–stone. Each player holds one hand behind their back and chooses either scissors, paper or stone by appropriate placement of the fingers: two separated fingers for scissors, flat palm for paper, clenched fist for stone. The payoffs (wins and losses) are governed by the rule that scissors cuts (beats) paper, paper wraps (beats) stone, and stone blunts (beats) scissors. Suppose Alice and Bob play, with Alice going first. With a point score of 1 for a win, –1 for a loss and 0 for a draw, the table of payoffs for Alice (technically called the payoff matrix) is as shown in Figure 58. The payoffs for the second player are the same, except 1 and –1 are swapped. That is, if Alice wins then Bob loses, and conversely.

Intuitively, scissors–paper–stone is fair: neither player has a clear advantage. This would not be the case if, to take an extreme example, the payoffs were such that Alice always wins, with 1's everywhere. In fact, scissors–paper–stone is symmetric: it treats both players equally. I won't formalise this idea of symmetry – it can be done, but it's technical and doesn't have a lot of content – but whatever move Alice chooses, Bob has a choice of one winning move, one losing move and one move that leads to a draw. So there is no bias against Bob. By the same argument, there is also no bias

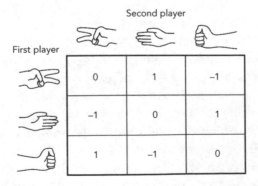

Fig 58   Table of wins (1), losses (–1) and draws (0) for the first player in scissors–paper–stone.

against Alice. We therefore expect that in the long run neither player will come out ahead to any significant degree. This turns out to be true, as long as the players don't introduce a degree of bias by making 'bad' choices.

Suppose, for instance, that Alice chooses scissors significantly more often than paper. Then Bob may notice. If he does, he could choose stone on every play and come out ahead, because in that case Bob wins whenever Alice chooses scissors, loses when she chooses paper and draws when she chooses stone. So Bob will win in the long run. In practice, if Bob's strategy were that obvious, then Alice would notice and start playing paper every time instead. However, the same reasoning applies if Bob makes random choices, but biases them in favour of stone: he will come out ahead if Alice favours scissors.

Pursuing this analysis leads to the reasonable conclusion (also a consequence of symmetry) that Alice should choose each possible move at random, with probability 1/3. Bob should do the same. Indeed, if one player departs from such a strategy, either by introducing regular patterns such as alternating paper with scissors, or by using probabilities that differ from 1/3, then the other player can find a response that wins in the long run.

Scissors beats paper beats stone beats scissors ... Familiar?

Orange beats blue beats yellow beats orange.

Could the common side-blotched lizards' mating games somehow be analogous to scissors–paper–stone? And if they were, what would that tell us?

* * * * * * * * * * * *

The great Hungarian-American mathematician John Von Neumann was one of the father figures of computing and a polymath who ranged over many areas of mathematics. A child prodigy born into a Jewish family, he spent four years teaching at the University of Berlin, and then went to Princeton University in the USA. When the Institute for Advanced Study at Princeton was created in 1933, he was one of the founding professors. Another was Albert Einstein. In 1927, having turned his mind to economics, Von Neumann invented a new branch of mathematics: game theory. A year later he made a fundamental discovery, the minimax theorem. Further developments led to the 1944 *Theory of Games and Economic*

Behavior, written with Oskar Morgenstern, which hit the front page of the *New York Times.*

A game, in Von Neumann's sense, is a simple mathematical model of two (or more) competing players, each faced with various choices, in which the payoff to each player depends on the combination of choices that they make. The players are assumed to know the table of payoffs, but have no knowledge of their opponents' choice. The game can be played just once, in which case we have to analyse the probability of winning or losing, or it can be played many times, in which case we can analyse the frequency of winning or losing (and how much is won or lost). A basic theorem in probability theory, the law of large numbers, says that in the long run the frequencies 'almost always' give the probabilities, so the two ways of thinking are mathematically equivalent. The usual choice is to consider what happens when the game is played many times, because our intuition for this is better than our intuition for one-off probabilities.

Scissors–paper–stone is a typical game, with one exceptional feature: its threefold symmetry. Most games treat different combinations of players in different ways. For instance, in the hawk–dove game, the players are in contention over some resource. Hawks always choose to fight, and escalate the battle until either they are injured or the other player breaks off the engagement. Doves always retreat from hawks. Depending on the entries in the payoff matrix, there can sometimes exist mixed strategies in which the best way to play is to switch randomly from hawk to dove and back again with particular probabilities.

Game theory first took off in 1928, when Von Neumann proved the minimax theorem. This states that in a particular class of two-person games with a very simple structure, there always exists a mixed strategy that permits both players simultaneously to make their maximum losses as small as possible. But this discovery was only the beginning. Another important piece of the puzzle fell into place when John Nash, the subject of the book and movie *A Beautiful Mind*, made a fundamental breakthrough for games with many players. He defined the concept of a Nash equilibrium and proved that one always exists. A set of players is in Nash equilibrium if each member of the set is making the decision that is best for them, given the decisions that the others have made. This is a sensible candidate for a rational strategy.

The person most responsible for the systematic application of game theory to evolutionary biology is John Maynard Smith. In 1973, in collaboration with the London-based American population geneticist George Price, he put forward one of the most important concepts in the field: that of an evolutionarily stable strategy. This is a refinement of a Nash equilibrium, and it pins down the conditions under which no mutant can successfully invade a population: a type of evolutionary stability.

Imagine a population of organisms, all of which have adopted – evolved – a particular survival strategy. In a genetic interpretation, this strategy will be inherent in their genes, as a result of many generations of natural selection. The organisms will not be consciously aware that they are adopting a strategy; it will simply be something that they do naturally, which has evolved because it works. Now suppose that there is some kind of genetic mutation, so that a similar organism, with a different strategy, suddenly appears in their midst. Can the mutant successfully establish a lineage of surviving descendants, or will it be weeded out by natural selection?

For example, consider the hawk–dove game in the trivial case where the population consists only of doves. This is not an evolutionarily stable strategy, because any hawk mutant can successfully invade – hawk always wins against dove. That is, hawk receives a positive payoff, while dove gets zero.

Maynard Smith devised a mathematical definition of an evolutionarily stable strategy. Suppose there is a finite list of available strategies. Let $E(A, B)$ be the payoff to an individual who adopts the original strategy A against an opponent who adopts strategy B. This is an entry in row A and column B of the payoff matrix.

Before the mutant appears, there is only one game in town: the entire population is playing the same *Old* strategy, and the payoff to each individual is $E(Old, Old)$. When the mutant appears, it adopts a strategy *New*. The payoff to the mutant is then $E(New, Old)$. If $E(Old, Old)$ is greater than $E(New, Old)$, then the mutant will lose the competition against any member of the original population, so its lineage will be weeded out. There are two other possibilities: either $E(New, Old)$ is greater than $E(Old, Old)$, or the two are equal. In the first case, the mutant wins and its lineage survives: it has successfully invaded the population. In the second case, the mutant loses out if the original strategy *Old* has a greater payoff against *New* than *New* does against itself; that is, if *E(Old, New)* is greater than *E(New, New)*.

The *Old* strategy is said to be evolutionarily stable if no mutant can successfully invade.

Some games have evolutionarily stable strategies, others do not. A general payoff matrix for two strategies *Old* and *New* looks like this:

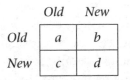

An evolutionarily stable strategy exists provided *a* is smaller than *c*, and *d* is smaller than *b*. The strategy concerned adopts *Old* with probability $(b-d)/(b+c-a-d)$, and *New* with probability $(c-a)/(b+c-a-d)$.[1]

When applying these models to real examples, the main difficulty is to estimate the entries in the payoff matrix. In principle, they could be estimated by playing one strategy against another many times and seeing what happens on average. But in practice this may not be possible. Suppose, for instance, that we are trying to understand some stage in the evolution of dinosaurs. We can't pit dinosaurs against one another and see who wins. So the entries have to be estimated on the basis of other factors.

Game theory sheds light on the evolution of new species, which can arise when changes in the environment render a single-species strategy evolutionarily unstable. If so, then a mutant can successfully invade – and given enough time, a suitable random mutation should arise. This doesn't explain speciation, but it does determine circumstances under which it might or might not be possible.

· · · · · · · · · · · ·

Darwin titled his book *The Origin of Species*. Biologists have built on its main ideas ever since. So can you guess what is one of the biggest mysteries in evolutionary biology today?

That's right. The origin of species.

However, that does not mean that Darwin was talking nonsense and species have not evolved. It reflects the difficulty of reconstructing fine details of processes that occurred millions or billions of years ago, and the rich complexity of today's living world. This is hardly surprising. What is surprising is how strong the evidence for evolution is, and how much we already know about it.

You may wonder how scientists can be sure that evolution has happened when they don't understand many of the details. However, we are faced with exactly this situation on a regular basis. We know that our child has learned things at school, but we weren't in attendance ourselves. We know that he or she has learned to speak, requiring changes in the child's brain, but we don't have high-resolution before and after brain scans to prove it. We know that the cat brought a mouse in last night, because the gruesome evidence is on the kitchen floor, but we never saw the cat do it. Science is seldom about direct observation: it is nearly always about indirect inference.

We know that evolution has occurred throughout the history of life on Earth because many independent lines of evidence attest to the general nature of the process. Some of this evidence has survived for millions of years. We can measure the sizes of fossil horses, correlate them with the ages of those fossils as determined by the geological strata in which they are found, and see a slow but steady trend towards larger and larger animals. But if we want proof that two particular horses were competing with each other at some particular instant of time, say 10.34 in the morning on 16 April 18,735,331 BC, then nothing short of a time machine would enable us to observe the competition directly. Instead, we infer that in general terms the horses were competing with each other, because it is very hard to see how they could have *avoided* competing. Population growth, if unchecked, would soon have caused horses to overrun the entire planet. So something must have checked it, and virtually any such process is a form of competition.

Were they competing for access to females? For food? What kind of food? Why didn't those two horses have enough for both of them? Which one won? When the required level of detail gets too high, there is no serious chance of answering such questions.

• • • • • • • • • • • • •

Not only are we unsure of many details of the evolution of species, but we don't really have a good definition of what a species *is*. Again, that does not mean that blackbirds cannot be distinguished from whales. But certain fine distinctions are difficult to pin down precisely. Ironically, this very difficulty supports the theory of evolution: if species are not always discrete, separate groups, then it is easier to see how natural selection might cause new species to split off from old ones.

You might think that 'species' must have a straightforward definition. After all, taxonomists classify organisms according to which species they belong to. In the Linnaean scheme, you and I belong to the species *Homo sapiens*, my cat belongs to the species *Felis catus* and the silver birch tree in the garden belongs to the species *Betula pendula*. This shows that particular species can be defined, but it no more tells us what a species is than the list *Toyota Avensis*, *Ford Mondeo*, *Volkswagen Golf* tells us what a model of car is.

One of the most popular definitions of species was advocated by the German-American ornithologist Ernst Mayr: species are groups of interbreeding natural populations that are reproductively isolated from other such groups. This definition applies only to sexual organisms, because 'interbreed' requires sex. As a working definition – a guideline that works fine most of the time – it is pretty good. However, it has a few drawbacks if it is taken literally and expected to apply in all cases.

For instance ... There is a chain of gulls, more or less continuous, that begins in Britain, goes right round the world, and ends up close to where it started. At one end are herring gulls; at the other, black-backed gulls. These two types of gull satisfy Mayr's definition: they do not interbreed, hence are 'reproductively isolated'. They look different, and they are different. They both coexist (without interbreeding) in Britain: mixed urban colonies are found in Bristol, Gloucester and Aberdeen. However, along the chain, each group of gulls can and does interbreed with its neighbours. So by that same definition, all neighbours belong to the same species as each other. Therefore herring gulls and black-backed gulls must also belong to the same species. But they don't. It's like a string of beads, each one the same colour as its immediate neighbours, but with black at one end and white at the other.

An enormous variety of alternative definitions of 'species' have been proposed over the years. Mayr's remains popular, but there are circumstances in which it seems inappropriate: the gull story is far from unique. Alternatives to interbreeding include the potential to exchange genetic material, genetic similarity, morphological similarity, ecological similarity, common ancestry and technical ideas in cladistics.

Massimo Pigliucci is a biologist at Lehman College in New York whose background includes genetics, botany, ecology and – unusually – philosophy. He analysed the different definitions of speciation in the literature, and found all of them lacking. Just as Mayr's proposal is confounded by an arc of gulls, so each of the others fails to match some aspect of the rich and messy reality of biology. On the other hand, each works pretty well within a limited domain and for specific purposes. Pragmatically, that might be considered good enough: this is the empirical view that 'species' is a convenient way to distinguish organisms, and criticisms of any particular definition are mostly linguistic nitpicking. But that doesn't answer a basic question: is 'species' a fundamental level of organisation of the biological world, or is it an artificial classification scheme foisted on us by taxonomists, with no real significance for actual organisms? Pigliucci puts it like this:[2]

> The so-called 'species problem' is one of those topics of discussion among evolutionary biologists that has been present since before Darwin's publication of the aptly titled *Origin of Species* (Darwin himself referred to it as an already old problem) and will probably never go away ... On the one hand, [biologists] tend to turn away in disgust when species concepts are brought up by colleagues, are the subject of papers, or are discussed at conferences. On the other hand, they simply cannot resist the temptation to offer graduate seminars on the topic and avidly read anything that is published on the subject.

Many biologists consider the whole issue to be merely one of semantics – finding an acceptable definition for practical purposes, which matches empirical observations in the field. Pigliucci argues that the problem goes deeper, with 'strong philosophical overtones'. He discusses three main themes that dominate the philosophical literature on the problem: critiques of definitions proposed by biologists, analysis of what kind of *thing* a species is (individual? group? natural type?) and the possibility that more than one concept of species is needed anyway, depending on context and purpose. The solution he proposes is founded on the philosopher Ludwig Wittgenstein's idea of 'family resemblance'. The family resemblance, he suggests, is real: it has biological significance, and is not just a human invention that provides neat lists of types of organism. However, by its nature it is difficult to pin down a simple, neat, tidy definition that characterises this kind of family resemblance.

Here is Pigliucci's main conclusion, rephrased in terms congenial to a mathematical biologist. First, we exploit the idea of multidimensional spaces to represent a list of phenotypic (and/or genotypic) data as a point in a conceptual space of many dimensions. Call this phenotypic space. Then we plot the points that correspond to each organism under investigation. We also need some notion of how far apart or close together two organisms are. There are many ways to set up such a 'metric' – by measuring the differences in characters such as wingspan or beak size, by comparing gene sequences, by looking at patterns of behaviour – what food do they eat? – and so on. A cluster is then a collection of data points in phenotypic space whose members are closer to one another than they are to anything else. What emerges from this viewpoint is that a satisfactory definition of species is elusive *because* a definition of cluster is elusive – it depends on the choice of metric, for instance. That's Pigliucci's philosophical point. The distribution of organisms in phenotypic space is not a convention: it is real, and can be observed. The problem is how to break the distribution into clusters, and what that is taken to mean. On the whole the answer is usually fairly obvious, which is why all the traditional definitions work pretty well, most of the time. When they don't, there are two ways to proceed: tinker with the metric, or refine the definition of a cluster.

A key feature of this proposal, says Pigliucci, is that it stops biologists 'wasting their time by trying to empirically solve a problem that has philosophical components that cannot be settled by the accumulation of new data'. Collecting vast quantities of data about, say, the disputed taxonomy of lizards living on islands down the west coast of North America will never resolve the taxonomic issues, which are really about the choice of metric and the concept of cluster involved. However, such data may contribute to the analysis of the situation when the metric and concept are chosen, and could even guide their choice.

• • • • • • • • • • • •

Looking for clusters within data may sound vague and woolly, but there is an entire branch of statistics devoted to the topic: it is called cluster analysis. As with any well-developed branch of mathematics, cluster analysis applies many different methods to the same general class of problems, and which one works best depends on the problem. I'll concentrate on the simplest method. In practice it is performed using algebraic calculations on numerical data, but the underlying idea is easier to understand in visual form.

Suppose that a field ornithologist on some exotic island is observing birds. One thing they might do is capture birds, measure various characters and record the measurements. For efficiency, they would probably record several dozen characters, but to keep the explanation simple and make it possible to draw pictures, I'll consider just two: beak size and wingspan. It doesn't actually matter what these variables represent: just that for each individual bird the ornithologist gets two numbers. Having collected the data, the ornithologist plots them on a diagram, which might resemble Figure 59, where I've used made-up data for illustrative purposes. I've also omitted scales from the axes.

The left-hand graph shows the plotted data. It's difficult not to notice that the points form two distinct clusters. The clusters might correspond to two distinct species, or perhaps two subspecies within a given species. Which of these is appropriate depends on the level of detail: how wide the square is relative to the numbers that constitute the data. If it's a big square, then birds in one cluster are significantly different from those in the other cluster, and the clusters represent species. If the differences in data are just a few per cent, we might be talking about subspecies.

As well as the two clusters, a single point lies on its own at

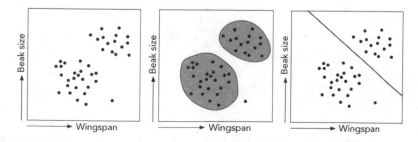

Fig 59 *Left*: Data. *Middle*: Two clusters and an outlier. *Right*: Separating the clusters.

lower right, and it's not clear whether it belongs to the cluster to its left, or is part of a third cluster. Such points are known as outliers because they are rare exceptions to the overall pattern. The method of data analysis means that their effect is small, and it makes little difference to the results if they are discarded. But this point might also represent a new, rare species, so in practice it would be better to collect more data.

The eye easily separates the data into two clusters, but it is not straightforward to program a computer to perform such a task. The simplest form of cluster analysis seeks to separate the data into two subsets by, in effect, drawing a line between them. More precisely, the method sets up some combination of the two variables, such as

$$0.5 \times (\text{wingspan}) + 7.3 \times (\text{beaksize})$$

together with a threshold value, say 15. Any choice of these three numbers splits the data into two subsets: one consisting of data for which the combined value is greater than the threshold, the other consisting of data for which it is smaller. If these numbers are chosen correctly, then the two subsets will be widely separated, but within each separate subset the numbers will be much closer together. All of this can be made precise, and carried out as a numerical calculation.

Figure 60 shows real data for the horned beetle *Onthophagus nigriventris*. The two variables plotted are body size (horizontal axis)

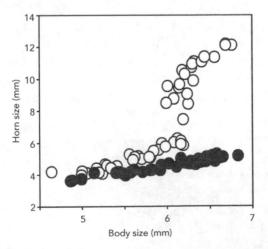

Fig 60 How the beetle changed its horns. Relation between body size and horn length for males (open circles) and females (solid circles) in the horned beetle *Onthophagus nigriventris*.

and length of horn (vertical axis). The open circles are males, the black ones females. Ignoring this distinction, the most obvious clusters are all the females together with some of the males, and the rest of the males. The two are clearly separated – for example by a horizontal line drawn through horn length 7 mm.

If we introduce a third variable, sex, plotted in a third dimension, then the males and females are immediately separated, making the point that extra data measuring new phenotypic variables can dramatically change the clusters. Since the distinction between males and females is standard, and we expect significant phenotypic differences between the sexes, it is a reflex to separate the data by sex – hence the two colours of dots.

The males here all belong to the same species, even though they are split between two clusters. In particular, they can all interbreed with the same females, and the females form one cluster. Nonetheless, the males clearly do split into two clusters, and recording extra variables can only reveal further splits, if there are any. These clusters represent two different morphs – phenotypes that have evolved according to two distinct strategies within the same population. Some males, called majors, go for big horns and an aggressive personality. They employ their large horns to fight other males, for access to females. The evolutionary advantage of this strategy is clear: the bigger your horn, the more formidable you are in combat. The other males, minors, develop rudimentary horns and avoid combat. Experiments suggest that one advantage of this strategy is greater manoeuvrability in tunnels. But it might allow the minors to sneak up on females while the majors are having their fights, because minors resemble females a lot more than they do majors.

This kind of separation of mating strategies occurs in the males of many species; we've already seen it in lizards. It is particularly extreme in horned beetles. It might be a sign that the species is in the process of splitting.

The use of straight lines, or their equivalents when there are more variables, avoids the trap of using ever more complicated formulas to model data. This can lead to an almost perfect fit between data and model, but one that is entirely meaningless. However, sometimes there are genuine clusters, but no straight line can separate them. Think of a tight circular cluster surrounded by a larger horseshoe-shaped one, for instance. Clusters like these can be detected by allowing nonlinear terms in the algebra – such as the square of the beak size or the beak size times the wingspan.

· · · · · · · · · · · · ·

Recall that the divergence of a single species into two (or perhaps more) distinct ones is called 'speciation'. Evolutionary biologists recognise many types of speciation, but there are two main ones: allopatric and sympatric. The words are derived from Greek: *allos*=other, *sym*=together, *patra*=homeland. Speciation is allopatric if it takes place in different locations, in a specific sense that I will explain in a moment, and sympatric if it takes place in the same location – also in a specific sense.

The big issue in speciation is not the *potential* for a group of more or less identical, happily interbreeding organisms to diverge. Well before Darwin, everyone knew that descendants are not identical to their ancestors, and the breeding of domestic animals showed that the same kinds of ancestor can give rise to very different descendants. For instance, a breed of sheep with short wool might occasionally have offspring with long wool, or wool of a different colour. In this instance, human intervention can persuade the sheep population to realise that potential, but the new types of sheep are not new species, just new breeds. Nevertheless, the potential to diverge must have been present. For true speciation to occur, whether by human hand or by unaided nature, the problem is to keep the new types separate. Otherwise they may interbreed, and then they are likely to reconstitute the original stock.

Somehow, the two diverging groups must be *reproductively isolated*.

Traditional methods of animal breeding achieve this by direct intervention: the breeders control which animals mate with which. This is how breeds of pedigree dog are maintained, and one of the reasons why they are so expensive. Distinct breeds of dog, left to their own devices, would revert to a population of mongrels within a few generations.

Allopatric speciation achieves the same result through some form of geographical isolation ('different homeland'). The idea is that some natural feature, such as a river, a mountain range or a land bridge, separates what was initially a single population into two distinct ones. Once separated, the two groups can change, and they are likely to do so in different ways because they have ceased to interact with each other. If this process continues for long enough, it may become impossible for members of the two groups to interbreed. At that point, the groups constitute distinct species.

Here's a classic example. The Caribbean Sea just north of the Isthmus of Panama, and the Pacific Ocean to its south, contain many organisms that are closely related but constitute distinct species. Working from the Scripps Institution of Oceanography in San Diego, the marine biologist Nancy Knowlton has studied snapping shrimp.[3] Very similar species of shrimp – they look almost identical – are found on both sides of the isthmus, but when brought together they do not interbreed. Every single one of seven distinct lineages of snapping shrimp has split into two species (in three cases there is also a third subspecies): one branch lives in the Caribbean, the other in the Pacific (see Figure 61).

It is straightforward to provide an evolutionary explanation, and difficult to find anything else that makes sense of this remarkable pattern. We know independently from geological evidence that

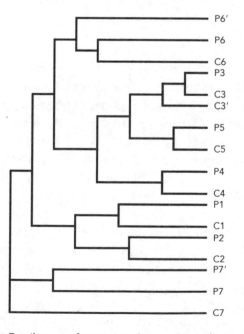

Fig 61   Family tree of snapping shrimp. Seven lineages, four pairs of species and three triples (including subspecies). P=Pacific, C=Caribbean. One species in each pair lives either side of the Isthmus of Panama.

three million years ago falling sea levels and rising land filled in the gap between the two halves of the American continent, creating the Isthmus of Panama and separating the Pacific Ocean from the Caribbean and the rest of the Atlantic. Before the land bridge formed, each lineage constituted a single species. Afterwards, allopatric speciation caused each lineage to split into two (or more) species, the ones we have today. For each lineage, the modern species of snapping shrimp are descendants of the same ancient species, which 'drifted apart' genetically once the two populations were prevented from interbreeding by an impassable land barrier.

This evolutionary hypothesis makes a quantitative prediction. If we can work out the time that has elapsed since the species began to diverge, independently of the geology, we should obtain a result of about three million years. The discovery of DNA has made such estimates possible, because mutations occur at an approximately constant rate.[4] If the two times disagree, something is wrong. As it happens, they do agree.[5] So here we have a specific prediction made by evolutionary theory and confirmed by experiment.

There are more ways to test the theory of evolution than watching a cauliflower and insisting that it must change into a cat before your very eyes.

. . . .

The allopatric mechanism for speciation is simple and direct; computer scientists would call it WYSIWYG – what you see is what you get. To separate something into two parts when it wants to stick together, drive a wedge through it. Perhaps for this reason, most biologists believe that most speciation events have been allopatric, and they have found examples everywhere – the African and Indian elephant, squirrels on the two sides of the Grand Canyon, the Faeroe Island house mouse ...

As a mathematician, I tend to be suspicious of WYSIWYG explanations. They put in what they want to get out, which smacks of circular logic. The world is usually less direct. Sympatric speciation is definitely less direct, but it is also more puzzling. For a long time, it was thought to be exceedingly rare, if not impossible. Initially there is a single species, a group of more or less identical creatures (except for sex if the species is sexual). They are all able to interbreed with one another, with fertile offspring. They are all conveniently located in the same place, so there is no lack of opportunity to interbreed. Suppose that for some reason this single group starts to split into two or more genetically different types. Then there seem to be two immediate reasons why this split should not develop into a lasting division into two distinct, non-interbreeding groups, let alone two groups that *cannot* interbreed with fertile offspring – two distinct species.

One reason is genetic. In the early stages of the split, the two new types can interbreed and there is no geographical barrier to stop them. Because the new types are small in number but the main population is large, the mates of the new types will almost always be members of the main population. But then, the new genes will be overwhelmed by the existing ones. So as soon as a group begins to acquire genetic differences, those changes will be snuffed out – swamped by the main gene pool – and the result recreates the genetic make-up of the original species.

This is the problem of 'gene flow'. It presents a stabilising force that mitigates against sympatric speciation.

Another objection is evolutionary. At least one of the new types differs from the original population. In order for this new type to evolve, creatures of that type must be fitter than the original population. But if belonging to the new type makes some particular creature fitter, then the same must apply to all the others. So why don't they all change in the same way, thereby sticking together as a single species? The species might drift, as a whole, but it shouldn't *split*.

I've already explained why the concept of 'fitness' needs to be treated with care, but the argument just outlined applies whatever specific meaning is attached to that term. It seems watertight. So it looks as though sympatric speciation is impossible. And that's what most biologists thought until the last decade or so. Then a series of theoretical models and observations, mostly in laboratories but occasionally in the wild, made some of them rethink the whole question. And it has turned out that the arguments against sympatric speciation are not as strong as they appeared to be. Agreed, *something* has to stop gene flow from gluing the nascent split back together before it has really got going. But geographic isolation, as in the allopatric case, is not the only game in town. It is not necessary for organisms to be physically prevented from interbreeding. It is enough that, for some reason, they don't.

. . . .

A case in point is the (I say 'the', but see how the story goes) African elephant. When I was at school, we were taught that there are two species of elephant, African and Indian. Almost all taxonomists were happy with that, but for a century or so a few mavericks kept wondering about the forest and savannah elephants in Africa. No elephant can be called sylph-like, but forest elephants are significantly slimmer than savannah ones, and there are other differences in form and behaviour that suggested to these taxonomists that there must actually be *two* species of African elephant: forest and savannah. Nonsense, said the rest: the forests are adjacent to the savannahs, so the elephants in the forest can interbreed with those on the plains, and gene flow will do the rest. They may be distinct subspecies, but they can't be different species.

The argument raged for a century, all of it inconclusive. Then, in 2001, *Science* reported that a DNA identification system, set up to trace poached ivory, showed that African elephants consist of two different species.[6] The researchers were expecting slight variations between the genetics of forest and savannah elephants, consistent with their being subspecies, but the difference was much greater than expected. The DNA evidence showed that the species diverged about 2.5 million years ago. In fact, the genetic difference between the African forest and savannah elephants is 58% of that between either of them and the Indian elephant. So now most taxonomists accept that there are two elephant species in Africa: the forest elephant *Loxodonta cyclotis* and the savannah (or bush) elephant *L. africana*.

Why doesn't gene flow reunite the populations into a single species, then? Although the forest is adjacent to the savannah, and it is true that elephants from the forest can mate with those from the plains, they seldom do. One reason is obvious: they don't get much opportunity. A female and a male must come together accidentally at the forest edge, just when both are ready to mate – which in elephants is a fairly small proportion of the time. Even if they manage that, *she* has to fancy *him*, and often she doesn't. So although their geographical ranges overlap, or at least abut, gene flow does not glue the two species back together.

Some taxonomists argue that the African elephant story is still allopatric speciation: in effect, the edge of the forest acts as a geographical boundary. But while they were arguing that there had to be just one species, thanks to gene flow disrupting sympatric speciation, this 'boundary' never figured in their arguments. Yaneer Bar-Yam has devised mathematical models based on genetic diffusion which show that gene flow can be prevented without having an impenetrable barrier.[7] A sparsely spaced series of obstacles is enough.

We don't know exactly how the two species of African elephant diverged. But if the distinction between allopatric and sympatric speciation means anything (and those who believed sympatric speciation to be impossible certainly thought it did), then the African elephants diverged sympatrically. Whenever a species starts to diverge, there will always be *some* sense in which one group is different from the other, otherwise there is no divergence. That difference may well affect the prospects of members of the two groups mating: not whether they *can* do so, but how likely it is that they *will*. On the other hand, whenever a species starts to diverge, the creatures will initially be in much the same location. So seizing upon some tiny difference and arguing that it constitutes allopatric speciation is really just renaming sympatric speciation.

The divergence of species involves the fate of individuals and their descendants, as well as family groups and the population as a whole. You don't just wake up one morning to find two neatly separated groups of elephants, when the night before there was only one. From Massimo Pigliucci we have learned that 'species' is best seen as a particular scale of clustering in phenotypic space; in the same way, 'speciation' is a divergence of clusters on a similar scale. It starts with one cluster and ends with two, but could be very complex and messy in between. Mathematical models suggest that this may well be the case.

So sympatric and allopatric speciation are convenient broad categories, not mutually exclusive alternatives. As such they are useful because they capture a key difference: how the animals do or do not interact during the speciation process. Are they all in one place, or not? And the current view is that they may well be in the same place, so the speciation can be sympatric, while a host of influences can disrupt gene flow and allow the genetic and phenotypic split to grow. As the American geneticist and complexity scientist Stuart Kauffman put it at a conference in Sweden some years ago: the key step to speciation is getting a foot in the door. If that's possible, then the door can be kept open, and maybe widened.

. . . .

Darwin's discussion of evolution is purely verbal, aside from one quasi-mathematical diagram, a tree-like figure showing how repeated small-scale branchings can combine to give large-scale divergence. But because evolution is a complex and sophisticated process, words alone are no longer adequate to describe it or debate it. Increasingly, the subject is being studied using mathematical models. The advantage of such models is that they make the assumptions involved clear. The disadvantage is that no model can capture the full complexity and vast scale of evolution, which has been going on in parallel across the entire planet for nearly four billion years.

Traditionally, biologists used that complexity as a reason to ignore mathematics and fall back on words. But verbal descriptions are even less able to capture the complexity of evolution than are mathematical models. Worse, they are imprecise and open to misunderstandings and ambiguities. The models clarify the concepts, the assumptions and the relations between them. That is what models are for. A model that describes every detail *exactly* would be like a map of the world that is the same size as the world.

The argument against the possibility of sympatric speciation has flaws other than the ones we've just encountered. In particular, one of the basic assumptions involved turns out to be wrong. This can be established by setting up specific mathematical models of sympatric speciation and investigating their implications; this has been done, and from several points of view.

A typical example is a paper written by the California-based scientists Alexey and Fyodor Kondrashov and published in the journal *Nature* in 1999, about a scenario that facilitates sympatric speciation.[8] They begin by observing that speciation models in which there is a mutation in only one gene (more properly, in only one genetic locus, a place in the genome where the gene resides) 'have very peculiar properties' because they are too simple to be realistic. It is therefore important to consider mutations that happen at similar times in two or more genetic loci, in what are known as multi-locus models. One scenario that leads rather directly to speciation arises when one gene confers a new character, and the other encourages mating patterns in which that character reinforces its own occurrence by a process known as assortative mating. For example, we saw in Chapter 7 that a single mutation in lacewings can change their colour from light green to dark green. The effect of predation now leads to more of the dark green insects being found in one type of environment: conifers, with dark green foliage. On light green grass, there will be more light green lacewings.

This is not yet speciation, because interbreeding between light and dark varieties remains possible and gene flow can remove the mutant. But suppose there is now a second mutation, causing light green females to prefer light green males, and ditto for dark green. Now, although the two varieties *could* interbreed, they don't. This sets the stage for further mutations, which now occur independently in the two varieties, causing them to drift further apart genetically. Eventually, even if they do happen to interbreed, the resulting hybrids may not be viable. Now the species have separated.

This example is somewhat artificial, because the new character does two things at once: it makes the lacewing fitter when it is in a particular environment, and it is also the character that females prefer. The Kondrashovs analysed a more realistic model in which one character affected the organism's fitness, but a different one affected mate choice. They showed that, again, there are circumstances in which the result can be sympatric speciation. Their model was a probabilistic one, analysing how the probabilities of particular mutations affected the frequencies of the corresponding phenotypes.

In the same issue of *Nature*, Ulf Dieckmann (International Institute for Applied Systems Analysis, Austria) and Michael Doebeli (University of British Columbia) found a different combination of genetic changes that can cause sympatric speciation.[9] Again, one of the characters involved was something affecting the fitness of the organism in a given environment, but this time the other was not a mating preference as such, but an 'ecological' character – one that affects the probability of mating occurring indirectly, according to the environment. In the case of the lacewings, the female need not possess a specific genetic preference, for instance: instead, the opportunities for mating arise when both males and females have

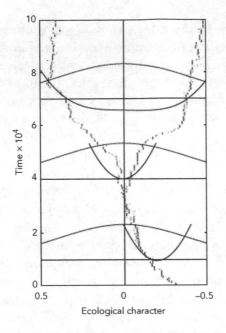

Fig 62 How a single species branches into two in the Dieckmann–Doebeli model. The three sets of curves are fitness functions. The scales are in arbitrary units used in the simulation.

the same colour. The light green ones are more common on grass, so light green females encounter more light green males; similarly for dark green insects on conifers. So this time it's not whom you prefer, but whom you commonly meet, that matters.

The end result is identical: assortative mating can occur, opening the door to sympatric speciation.

Dieckmann and Doebeli's mathematical model is different: it belongs to a class of models known as adaptive dynamics, and uses a differential equation rather than probabilities. That is, we write down equations that govern the rate of change of the sizes of the original population and the mutant population, and then work out the dynamics of the resulting system. Figure 62 shows a typical instance of speciation in a numerical simulation of this model.

• • • • • • • • • • • •

There is also a more general way to spot the flaw in the argument that sympatric speciation is impossible: to view speciation as an example of symmetry breaking. We met this idea in the previous chapter, in connection with the markings on animals. What has symmetry breaking to do with speciation? Are species symmetric? Well ... yes. But not in the way that stripes on a tiger are symmetric.

The same mathematical concept can be realised in many different ways. The interpretation of the symmetries is very different in the applications to markings and speciation. For markings, the relevant symmetries are rigid motions; for speciation the symmetries 'shuffle' the organisms like a pack of cards. Only in the abstract do the two applications share the same underlying mathematics. This, in fact, is where mathematics gets much of its power: by using the same idea in different contexts.

We saw that a symmetry is a transformation that preserves structure. Here, the transformations are permutations – shufflings of labels employed in the model to identify the individual organisms. If ten identical finches are conceptually labelled 1–10 in some order, then the mathematical model of their interactions should not depend on the choice of labelling. All possible permutations of the numbers 1–10 should lead to the same model – a severe constraint on its mathematical form. On the other hand, if five finches with small beaks are labelled 1–5, while another five with big beaks are labelled 6–10, then the mathematical description should permit relabelling within each group. But it should not, for example, allow label 5 to be swapped for label 6.

From this point of view, sympatric speciation is a form of symmetry breaking. If a population of nominally identical birds evolves into two distinct groups, then the resulting system has lost some of its symmetry (such as the transformation that swaps labels 5 and 6). For several decades, mathematicians and physicists have been developing a general theory of symmetry breaking, and it is now being applied to models of speciation. One of its most striking features is that there are many 'universal' phenomena that do not depend on specific details of the models, only on what the symmetries are and which of them is being broken. This theory can be applied to idealised models of speciation, where it leads to at least three universal phenomena.

The first is that when a population first speciates, it usually splits into precisely two distinguishable types (Figure 63, see over). Splitting into three or more types is rare, and a mainly transitory phenomenon. The second is that the split occurs very rapidly –

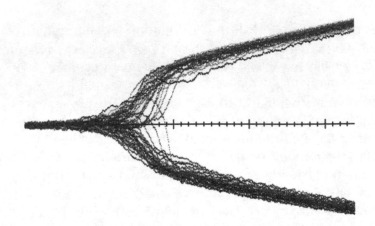

Fig 63 Simulation of sympatric speciation in a model with 50 clumps of organisms. Time runs horizontally, phenotype vertically. As time passes, a single species splits into two.

much faster than the usual speed of phenotypic change in the population. The third is that the two clumps will evolve in opposite directions: if one clump evolves larger beaks, then the other clump will evolve smaller beaks.

The symmetry-breaking models indicate that a key step in sympatric speciation is the onset of instability – exactly as in the game-theoretic models discussed earlier. As the environment or population size changes, the single-species state may cease to be stable, so that small, random disturbances can cause big changes. Like a stick being bent by stronger and stronger forces, something suddenly gives and the stick snaps in two. Why? Because the two-part state is stable, whereas one overstressed stick is not.

A population of organisms is stable if small changes in form or behaviour tend to be damped out; it is unstable if they grow explosively. Theory shows that gradual changes in environment or population pressure can suddenly trigger a change from a stable state to an unstable one.

There are two main forces that act on populations. Gene flow from interbreeding tends to keep them together as a single species. Natural selection, in contrast, is double-edged. Sometimes it keeps the species together, because collectively they adapt better to their environment if they all use the same strategy. But sometimes it levers them apart, because several distinct survival strategies can exploit the environment more effectively than one. In the second case, the fate of the organisms depends on which force wins. If gene flow wins, we get one species. If natural selection *against* a uniform strategy wins, we get two. A changing environment can change the balance of these forces, with dramatic results.

Specific mathematical models, such as those built by the Kondrashovs or Dieckmann and Doebeli, support this scenario. They provide a variety of biological mechanisms that can make the single-species state unstable and cause the symmetry to break. Intuitively, we can understand the common features of such models in simple biological terms, and suggest a genetic interpretation.

Imagine, for example, a species of finches with medium-sized beaks, all feeding on seeds from the same plant. Nominally they are all identical – that is, any differences are superficial and do not really alter the behaviour of the group. Their population expands until it is limited by the supply of those particular seeds.

Within this species, there will be a range of genetic variation, say in beak size. If the size preferred by natural selection is in the middle range, then gene flow beats environmental disruption, and the finches remain one species. The population size is well adapted to the food supply, so nothing much changes. But now suppose that a change of climate, say, reduces the food supply. Now there are advantages in having beaks that are not medium-sized, more suited to other types of seed.

A species can be thought of as a cluster in phenotypic space, so it is spread out, and not just a single point. So some birds will have slightly larger beaks than average, and others slightly smaller beaks. This is unavoidable: the average is in the middle. Birds with slightly larger beaks can feed on larger seeds. They then cease to be competition for the birds that have slightly smaller beaks. Once the balance swings in favour of avoiding the middle ground, the collective dynamics rapidly drives the birds into two distinct types. These types do not compete directly for food: instead, they avoid competition by exploiting seeds of different sizes.

As Kauffman said, once diversity has its toe in the evolutionary doorway there are many ways to amplify the split. The most obvious factor of this kind is one we have already met: assortative mating. Organisms in a given group share similar habits, eat similar food, and therefore meet up more often than they do with members of the other group. So the big puzzle is the initial split – which need not be especially large or dramatic. At a later date either of these clumps may split again, as continuing changes to the environment change the availability of resources. A cascade of such splittings leads to what biologists call 'adaptive radiation', where many new species arise from one ancestral species over a relatively short period.

At the moment, this kind of modelling is in its infancy. Its main contribution is to show that sympatric speciation is reasonable and natural, and to focus attention on the role of instability as a mechanism for species diversification. With more biological realism, the nature of these instabilities can be better understood.

In the meantime, the symmetry-breaking approach puts speciation in a new light. A stick breaks because the large-scale forces that act on it are inconsistent with its structural integrity. Precisely *how* it breaks depends on very fine detail about which fibre gives way first and how the consequences cascade – but if it didn't happen one way, it would have to happen some other way. *Whatever the details*, the stick will break. Similarly, species diverge because of an unavoidable loss of stability. The actual sequence of events – which gene does what, and in what order – is less important than the context in which these events occur. An overstressed stick must break. An overstressed group of organisms must either die, or speciate.

Is there any evidence for this type of scenario? Not directly, because the timescale for the evolution of new species is too great. But there do seem to be relics of past speciation events that match the symmetry-breaking scenario very closely. A particularly informative and historically important case, which displays features consistent with various mathematical models of sympatric speciation, concerns what we now call Darwin's finches, in the Galápagos Islands.

. . . .

The Galápagos Islands form an archipelago, a group of islands, in the Pacific Ocean. They are situated at the equator, 1,000 kilometres west of the coast of Ecuador. The name comes from the Spanish word for 'tortoise', reflecting the presence of the celebrated giant tortoises. There were a quarter of a million of them when the islands were discovered in 1535; today they number around 15,000. The archipelago contains about ten large islands and dozens of smaller ones, all of volcanic origin.

The geology of the Galápagos Islands is unusual, and it has had a profound effect on the creatures that live there. For hundreds of millions of years the islands have followed a remarkable cycle: they rise from the ocean floor at the western edge of the archipelago, move slowly eastwards, sink beneath the waves, and eventually disappear beneath the western edge of Central America. So at any given moment, the oldest and most eroded islands are to be found in the eastern part of the archipelago, while the newest and most volcanically active are to the west.

Today we are used to the idea that continents can move, but fifty years ago it was controversial, and sixty years ago it was considered crazy. In 1912 the Berlin meteorologist Alfred Wegener took seriously a similarity that hundreds of people must have noticed: the east coast of South America and the west coast of Africa fit together like adjacent pieces of a jigsaw puzzle. He argued that this was evidence for 'continental drift'. On geological timescales, the continents are not fixed; instead, they move, *very* slowly, over the surface of the planet.

The applied mathematician Harold Jeffreys objected that no physical mechanism can exert the gigantic forces required to make continents plough through the ocean floor. This was entirely correct; nevertheless, by the 1960s it became clear that Wegener was right. The ocean floor moves along with the continents, like a giant conveyor belt. New floor forms along mid-ocean ridges as magma wells up from the Earth's mantle, cools and spreads sideways; old floor is 'subducted', sliding back down into the mantle at the edges of continents. Indeed, it is sometimes *pulled* down. As a result, the surface of the Earth is divided into eight major 'tectonic plates' and many smaller ones. These plates are effectively rigid, yet they can move in complex ways, driven by huge convection currents in the molten rock of the Earth's mantle. They interact along their common boundaries.

The Galápagos Islands are poised at the junction of three plates: the Cocos, Nazca and Pacific plates. This meeting point, known as the Galápagos Triple Junction, is geologically unusual because the plates do not meet in a simple Y shape. Instead, two much smaller 'microplates' seem to be trapped at the junction, and they spin in synchrony with each other like two adjacent gearwheels. The Canadian geologist J. Tuzo Wilson explained this curious behaviour in 1963, suggesting that beneath the islands there is a geological 'hot spot' where a huge plume of molten magma rises up through the mantle, breaks through the ocean crust and forms volcanic cones. A similar hot spot is thought to create the Hawaiian island chain, but those do not lie on a plate boundary. For at least 20 million years, the Galápagos hot spot has remained in much the same place, though it has wobbled a little; the ocean floor has drifted eastward across it, carried by the movement of the tectonic plates. Currently the Nazca plate is moving at a speed of 60 kilometres every million years, while the Cocos plate moves 80 kilometres every million years. These speeds might seem too small to matter, but continental drift would carry a Galápagos island to the mainland of South America in a mere 12 million years, very short by geological standards.

The Hawaiian islands seem to have formed one at a time, each volcano becoming dormant as it is carried away from the hot spot, but the Galápagos Islands are more complicated. Nearly all of them have been volcanically active in the last few hundred years, which geologically is a mere instant, and their periods of formation have overlapped substantially. Today the newest island in the archipelago is Fernandina, which is an active volcano.

. . . .

As a consequence of their isolation and continual tectonic turnover, the Galápagos Islands are a bit like the well-known axe, the exact same axe that my father gave me, though it has had three new heads and four new handles. This rapid turnover of land has made the flora and fauna of the Galápagos unlike any elsewhere on the planet. Darwin spent five weeks in the Galápagos. Walking over jagged fields of coal-black lava, he was struck by the realisation that this was new land, and that its outlandish creatures must be new arrivals. Much later, what he found there started to sink in, and it had a big effect on him. None of this (aside from a few general remarks from the second edition onwards) appears in the *Origin*, but his letters and notebooks show how important the Galápagos Islands were to his thinking.

Darwin was an obsessive collector, and he brought back a collection of dead birds from the Galápagos. He thought they were varieties of blackbirds, finches and 'gross-beaks'. The ornithologist John Gould looked at the specimens and told Darwin that they

were all finches – a dozen or so distinct species. They differed in
body size, in colouring and, especially, in the shapes and sizes of
their beaks. The differences were not huge, but enough to indicate
distinct species. Today they are known collectively as Darwin's
finches, a name dating from 1936, and we recognise 13 species in
the Galápagos[10] plus another on the Cocos Islands.

In *The Voyage of the Beagle*, based on his diary, Darwin wrote:

> The remaining land-birds form a most singular group of finches,
> related to each other in the structure of their beaks, short tails,
> form of body, and plumage: there are thirteen species, which
> Mr. Gould has divided into four sub-groups. All these species are
> peculiar to this archipelago; and so is the whole group, with the
> exception of one species ... The males of all, or certainly of the
> greater number, are jet black; and the females (with perhaps one
> or two exceptions) are brown. The most curious fact is the
> perfect gradation in the size of the beaks in the different species
> of *Geospiza*, from one as large as that of a hawfinch to that of a
> chaffinch, and ... even to that of a warbler ... Seeing this
> gradation and diversity of structure in one small, intimately
> related group of birds, one might really fancy that from an
> original paucity of birds in this archipelago, one species had
> been taken and modified for different ends.

Because Darwin uncharacteristically omitted to record on which
island he collected which specimen, he missed a smoking gun for
natural selection: the finch species are often different on different
islands. But we now know that Darwin's intuition was correct. The
closely related genetics of Darwin's finches show that they diverged
from a single ancestral group about five million years ago. (Perhaps
a few founders were blown to the islands from the Central
American mainland by a storm, but that's conjectural.)

The first systematic study of the genetics and behaviour of
Darwin's finches was made by David Lack. Working as a
schoolmaster in Devon and later as field ornithologist at Oxford, he
wrote two books with the title *Darwin's Finches*: a scholarly one in
1945, and a more popular account in 1961.[11] Lack visited the
Galápagos in 1938, and among other things, he measured the sizes
of various birds' beaks. In his first book he suggested that the
differences in size were recognition signals – birds could distinguish
their own species by looking at the beaks. So he viewed beak sizes
as an isolating mechanism, something that prevented gene flow

even when it would otherwise be possible. But by 1961 Lack had revised his opinion, and now saw the differences in size as evolutionary adaptations to different food sources. Later studies confirm that view.

Lack's work has been continued by Peter and Rosemary Grant, husband and wife, both emeritus professors at Princeton. Since 1973 they have spent half of every year on the tiny Galápagos island of Daphne Major, capturing and releasing the birds, tagging them, measuring their sizes and shapes, and taking blood samples. Through their efforts we now know a great deal about Darwin's finches, their behaviour, form and genetics.

. . . .

One of the main implications of the symmetry-breaking model of sympatric speciation is striking: the two new phenotypes move away from the old one in opposite directions.[12] If, for example, a finch species with medium-sized beaks splits into two distinct species distinguished by beak size, then one will have bigger beaks and the other smaller beaks.

No such splitting has been observed directly, but that is to be expected because the evolutionary timescale is so long. Instead, we might hope to find modern traces of the evolutionary process. Fossil Darwin's finches would be lovely, but the Galápagos Islands are volcanic, and fossils hardly ever form in volcanic strata. However, there is another type of 'fossil' – the modern species themselves. These display a phenomenon known as character displacement: the phenotypes of distinct species *change* if both those species are present in a given environment. With a bit of imagination, you can view these changes as a kind of modern reconstruction of the likely evolution of the two species.

The species concerned are the medium ground finch *Geospiza fortis* and the small ground finch *G. fuliginosa*, henceforth *Medium* and *Small*. The character concerned is beak depth, the width of the beak at its base. *Medium* is found on the island of Los Hermanos (Crossman), but *Small* is not. Conversely, *Small* is found on Daphne, but *Medium* is not. However, both species coexist on Isabela (Albemarle).

When only one species is present, the beak depth is the same for either of them: the mean beak depth for *Medium* is very close to

10 mm on Los Hermanos, and the same holds for *Small* on Daphne. But when they coexist, they force each other apart: the mean beak depth for *Medium* on Isabela is very close to 12 mm, but that for *Small* is 8 mm. The average of 8 and 12 is 10: neatly consistent with the prediction made by the symmetry-breaking model of sympatric speciation. It is also worth noting that the other kinds of model used by the Kondrashovs and by Dieckmann and Doebeli also possess this 'constant mean' property.

This is character displacement, not speciation as such. But it can be argued that when the two species are placed in the same environment, we are reconstructing the original evolutionary competition ... and finding out what the result was. I don't claim that this observation is anything more significant than a straw in the wind, and grasping at straws is not always to be recommended. But it's intriguing to encounter exactly the predicted behaviour in the most famous example of speciation that there is – and the one that Darwin nearly missed.

15 Networking Opportunities

· ·

As we've seen when looking at the brain in Chapter 11, networks are hot property in biology and mathematics. They are hot property in physics and engineering too, and a popular buzzword in the business world. If you work in science or in commerce, it is very difficult not to run into networks. A ubiquitous example is the Internet, which by definition is a network of intercommunicating computers.

Networks abound in biology. We've already seen how the ability of nerve cells to network allows relatively simple components to produce rich and subtle types of behaviour. Indeed, if the 'connectionist' view of the human mind is anywhere near the truth, our ability to behave intelligently, our conscious awareness and our feeling (rightly or wrongly) that we have free will are all consequences of two things: the brain's intricate network of neurons, and its interactions with the body that contains it and with the external world.

In the case of the human nervous system, the network is a physical thing. Nerve cells are linked together by axons and dendrites, the body's hidden wiring. In most biological networks this linkage is more metaphorical. Ecologists study food webs in an ecosystem: which organisms feed on which. A food web is a network in which real organisms are 'linked' conceptually by the relationship of eating one another. Disease transmission can be thought of as a network in which individual people are linked by infection. A species can be thought of as a network of organisms, linked by everyday interactions such as competition for food or mates, and social behaviour.

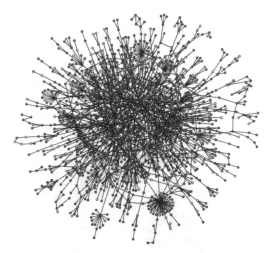

Fig 64 Network of protein interactions in yeast.

Some of the most important but least understood networks arise at the molecular level. We now know that biological development, in which a fertilised egg turns into an organism, is not just a matter of reading off information from DNA. Some parts of the genome, genes in the strict sense, provide instructions for making proteins. More precisely, they encode the order in which the constituent amino acids must be assembled: DNA doesn't perform the assembly. However, an organism is not just a bag of proteins: the right protein has to end up in the right place, and the entire system has to function as a living creature. One way to get protein to the right place is to make it there, and this involves switching the activity of a gene on or off depending on which cell it is in and where that cell is currently located in the organism. This switching is itself controlled, in part, by other genes, whose activity may be regulated by yet other genes. Collections of genes combine together to form genetic regulatory networks, and our understanding of how DNA affects development, and the day-to-day running of the body for that matter, depends on sorting out what these regulatory networks do and how they do it (see Figure 64).

• • • • • • • • • • • •

Social networking has become fashionable among humans, thanks to sites such as Facebook and Twitter. A real network, the Internet, helps us set up and maintain these social connections, but the social network is metaphorical. We were anticipated, millions of years ago, by an organism whose natural social behaviour sometimes creates real networks.

You wouldn't sign a contract with *Physarum polycephalum* for designing a railway system. For a start, it can't read or write: it is a species of slime mould. Its name means 'many-headed slime', and that's what it's often called. It can't compete in the intelligence stakes with humans, dolphins, octopuses or even mantis shrimp. There is dumb, there is dumber ... and there is slime mould.

However, this slime mould can indeed design railway systems. It does need a bit of prompting, but it does a pretty good job. In fact, as Atsushi Tero at Hokkaido University in Japan and a team of eight other researchers discovered early in 2010, *P. polycephalum* comes up with almost the same design for the Tokyo rail system that the engineers did.[1] And all they 'told' the slime mould was where the main cities are; the mould did the rest.

A few other scientists have persuaded slime moulds to solve other kinds of networking problems, such as finding their way through mazes. In principle, you could build a slime mould computer and do word processing on it, though it might take half the age of the universe to get as far as the first sentence. The slime mould's talents are not really suited to processing digital information at high speed. But when it comes to networks, slime mould is in its element.

Using slime mould to design a railway system is not as crazy as it sounds, because there are environmental conditions that cause slime moulds to structure their colony into a network of vein-like tubes, and to transport vital fluids along those tubes. A train system transports people along railway lines – it's actually a very similar problem. As well as using the slime mould to solve a problem about networks, the Japanese team used the mathematics of networks to find out how the slime mould did it, neatly closing the conceptual circle. The result might even be useful for human designers; not by using the slime mould to solve real design problems, but to simulate its strategies on a powerful computer.

P. polycephalum is a single-celled organism, a bit like an amoeba, except that this cell contains a large number of nuclei. Like an amoeba it can extend protuberances, which it uses to explore its surroundings – in effect, to go hunting for food. It spreads over dead logs or rotting leaves in a slimy yellow carpet, and the hunting goes on along the edge of the carpet, where individual 'plasmodia', regions that contain a single nucleus, seek food. Inside this exploratory boundary the organism forms itself into a network of tubes, reminiscent of the veins on the back of the human hand; just like veins, these tubes carry fluid, but the fluid is the protoplasm that makes up the interior of the cell. The protoplasm carries particles of food and other vital molecules as it flows, distributing the bounty to the entire organism.

Bizarre as this lifestyle may seem, it's an effective way to make a living, and slime moulds are very common. If you want to harness the networking abilities of slime mould, you have to present your computation to them in a form that appeals. And what appeals to a slime mould is *food*.

In its experiments, the Japanese team placed food sources on a flat surface at locations corresponding to the 36 main cities in the region. Think of it as a map of the area around Tokyo, with blobs of food for cities, but with no connecting roads or railways marked, because those would prejudice the procedure. Then they set their slime mould loose, by introducing a plasmodium at Tokyo and allowing it to go hunting.

To begin with, the plasmodium didn't have a clue where the food was, so it spread all over the map, forming a flattish layer. But as time passed, the layer contracted to form a network of tubes, linking the food sources. To avoid biasing the results by starting from Tokyo, the researchers performed another experiment, but this time they started with the mould spread all over the map. The results were very similar. To make sure that the mould formed a network, rather than just remaining spread out, they allowed any excess to spill over onto a large food source outside the area of the map.

The resulting network was very like the actual rail network connecting these cities (Figure 65, see over). The resemblance got better when the slime mould was deterred from entering what in the real world were mountainous regions by a light being shone on the corresponding bits of the map (*P. polycephalum* tends to avoid light). In effect, the researchers used light intensity to simulate terrain. With or without this extra tweak, the slime's network and the real one performed very similarly, according to a number of different measures of efficiency, such as the ratio of benefits to costs.

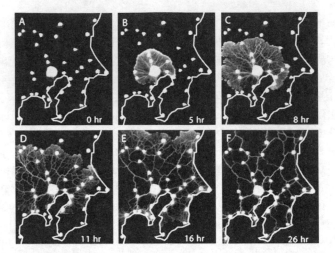

Fig 65 How the slime mould built its railway: six stages in the evolution of a biological network.

This behaviour is not entirely surprising, but there is a sign that something interesting is going on. The slime mould does not simply 'triangulate' the entire region by forming tubes between all neighbouring cities. Neither does the rail network. Both leave out potential links – and both leave out much the *same* links.

Is this similarity between the two networks a superficial accident, or a sign of shared origins? It's a bit of both. It might seem that the Tokyo network was 'designed', while the slime mould one evolved. Engineers worked out where to run the railway lines; the slime mould modified its network by expanding some tubes and contracting or eliminating others. But the real rail network also evolved: as cities appeared and grew, the rail system grew with it, adding new links. As the numbers of passengers increased, more lines and more trains were constructed, 'strengthening' the links in the network. Services that did not attract customers were abandoned.

If the engineers had been starting from scratch, with all cities in place but no rail network, they could have adopted a more rational and more global approach, designing the entire network to optimise whatever quantities seemed appropriate, such as transporting the required number of people cheaply and quickly. The slime mould could not have taken this approach. However, the engineers might still learn some useful tricks from the slime mould, as the Japanese team discovered when they devised a mathematical model of what the slime mould was actually doing.

Their model starts with a very fine random mesh of thin tubes, resembling the initial layer where the slime mould spreads all over the map. They wrote down simple equations for the amount of fluid that could be pumped through each tube, based on standard ideas from fluid dynamics. Their equations state that the amount of fluid flowing through a tube is proportional to the difference in pressure between its two ends, to its 'conductivity' (a measure of how big the tube is, proportional to the fourth power of its radius) and inversely proportional to the length.

To modify the sizes of the tubes, choose two random cities. Pump in extra fluid at one of them and extract it at the other, so that the total amount of fluid stays unchanged. Calculate the amount of fluid passing along each tube, using the equations. Make small changes in the diameters of the tubes, so that the ones carrying large amounts of fluid get bigger, and the ones carrying small amounts of fluid get smaller. Work out whether the change improves the efficiency of the network. If so, keep it; if not, try another random change. The exact method for doing this allows tubes to contract completely, attaining zero diameter: when that happens, they disappear from the network. Repeat this procedure over and over again, with new random choices of the two cities, and keep going until the structure settles down to something that hardly changes from one stage to the next.

Efficiency can be measured in many different ways: how much material can be transported, how rapidly it can be transported, how much benefit is gained for given cost. This evolutionary process can be deliberately tweaked to enhance any desired measure of efficiency: it is just a matter of selecting appropriate rules for how fast the tubes change in response to the quantity of fluid they are transporting. The team found that one tweak of this kind led to a network that in some ways outperformed both slime mould and the actual Tokyo rail system: it had the same transport efficiency but a better benefit-to-cost ratio. However, this network was more fragile: its ability to transport fluid (or people) declined significantly if parts of it were damaged or removed, whereas the slime mould network and the rail system are more robust.

There are other mathematical techniques for linking cities in a network, and the team compared their results with these as well. In cost terms alone, the most efficient network is a 'Steiner spanning tree', which has branches but no closed loops; in such an arrangement, whenever a branch splits at a Y-shaped junction, the angles in the Y are all 120°. This is the network that uses the shortest length of rail. But it is not a terribly good way to transport people, or slime mould protoplasm, because the link between two nodes can go all round the houses – from a distant city into the centre of Tokyo and out again, for instance. So travel time can be unnecessarily large. Neither the rail network nor the slime mould one looks remotely like a Steiner spanning tree.

Tero and his colleagues sum up their results like this:

> Our biologically inspired mathematical model can capture the basic dynamics of network adaptability through iteration of local rules and produces solutions with properties comparable to or better than those of real-world infrastructure networks. Furthermore, the model has a number of tunable parameters that allow adjustment of the benefit/cost ratio to increase specific features, such as fault tolerance or transport efficiency, while keeping costs low. Such a model may provide a useful starting point to improve routing protocols and topology control for self-organized networks such as remote sensor arrays, mobile ad hoc networks, or wireless mesh networks.

• • • • • • • • • • • •

Networks entered mathematics through a puzzle. In 1735 the prolific Leonhard Euler turned his mind to a topic of conversation among the good folk of Königsberg, then a city in Prussia, now Kaliningrad in Russia. The city was located on both banks of the River Pregel, and boasted seven bridges. These linked two islands to the banks, and to each other (see Figure 66). The burning issue was this: is it possible to take a walk that crosses each bridge exactly once?

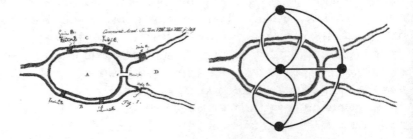

Fig 66 Problem of the Künigsberg bridges. *Left:* Euler's original puzzle. *Right:* Turning the puzzle into a network.

Euler didn't find a solution. He did something more difficult – he proved that *no solution exists*. His insight was to strip the problem down to its essentials. All that matters is how the land masses are connected. Their size and shape are irrelevant; worse, they get in the way of thinking about the problem. His actual argument was algebraic, assigning symbols to the land masses and the bridges, but soon afterwards it was reinterpreted graphically, a much more vivid way to represent the problem. This reduces the puzzle to a diagram in which dots are joined by lines, which I've superimposed on the map on the right of Figure 66.

In this diagram, each land mass – north bank, south bank and the two islands – corresponds to a dot, and we join two dots by a line whenever a bridge links the corresponding land masses. Altogether, we get four dots and seven lines. The puzzle now translates into a simple question: is there a path that includes each line exactly once? Euler discussed open paths, which start and end at different dots, and closed paths, which start and end in the same place. He proved that neither kind of path exists for the Königsberg bridge diagram. More generally, he characterised diagrams for which such paths do or do not exist.[2]

This kind of diagram was originally called a 'graph', but nowadays an increasingly common alternative is the more evocative and less ambiguous *network*. Euler's simple theorem about a puzzle was the first evidence for a broad principle: the architecture (or topology) of a network has a huge influence on what it can do.

Mathematically, a network consists of nodes (dots, vertices) linked together by edges (lines, links, connections, arrows). The nodes represent some kind of component, or agent, and two nodes are joined by an edge if and only if they interact. Edges may be bidirectional (interactions go both ways) or unidirectional (A influences B but not the other way round). The unidirectional case yields a directed network, whose edges are usually drawn as arrows. The edges may carry 'weights' to indicate the strength of the interaction. They may be of different kinds (fox predating on rabbit is different from rabbit predating on vegetation) or they may be nominally identical (fox A predating on rabbit X is near enough the same as fox B predating on rabbit Y).

Two graphs have the same architecture (or topology) if one graph can be obtained by rearranging the positions of the other one's nodes and edges, while maintaining the same connections

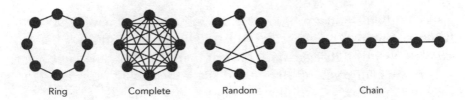

Ring   Complete   Random   Chain

Fig 67 Four types of network. Here all edges are bidirectional.

and arrow directions (and any additional decoration such as weights or types of edge). Important architectures include chains, rings, complete graphs and random graphs, as shown in Figure 67.

We can investigate networks in the abstract, in concrete mathematical models where the nodes and edges have additional structure, and in real networks with real agents and interactions. These three contexts are closely associated, but it is important to bear in mind that they are different. As long as we do that, we can safely employ the same words in all contexts – for example, talking of a rabbit as a node in a food web and drawing it as a dot.

Different people use networks for different purposes and ask different questions. An early pioneer was Stuart Kauffman, who employed binary switching circuits (each node can be either on or off) to model gene interactions in a cell. He found that the dynamics of the network depended critically on the average number of edges linked to each node. Innumerable types of network have been studied: discrete (cellular automata), continuous (differential equations), probabilistic (Markov chains), fractal (iterated function schemes). Many complex systems are networks. A radically new network architecture that has attracted attention is the *small world*: a regular network with near-neighbour connections, in which some edges are randomly rewired to become long-range connections, or some nodes become 'hubs' that are connected to unusually many other nodes.

There are several general theories of network structure and behaviour. Intensive work on statistical properties of random networks shows that as the probability of including any given edge increases, there is a transition at which most of the nodes suddenly link up into a single giant component. This result, metaphorically at least, has an application to epidemics. Here the nodes are people and edges indicate infection. The existence of a giant component shows that if the probability of disease transmission becomes sufficiently great, almost everyone will be exposed. Less obviously, it implies that there is a sharp threshold, below which the infection remains in numerous small 'pools' isolated from one another, and above which nearly everyone is exposed.

Another approach, introduced by the Japanese physicist Yoshiki Kuramoto, analyses network dynamics in which the effect of each node on those to which it is linked is small.[3] Think of an epidemic that is rarely infectious, for example. With these assumptions, useful quantitative predictions can be made about the behaviour of the network and its stability. More recently, theories have been devised to incorporate stronger connections and more exotic behaviour, including dynamical chaos.

• • • • • • • • • • • •

In 1996 Joanne Collier, Nicholas Monk, Philip Maini and Julian Lewis, a group of mathematical biologists at Oxford University, used a mathematical model to investigate a mysterious patterning process that had been observed in insects, nematode worms, chickens and frogs.[4] Cells can differentiate under the influence of suitable control genes and signals; that is, cells that originally have the potential to change into several different cell types choose one type and change into that. It is as if the genetic signals determine the cell's fate, and indeed this is the term that biologists often use. In some developing tissues, a mass of identical cells somehow differentiates into many types. Some apparently random collection of cells ends up with one fate, while the cells next to them have a different fate. The resulting cell types (there may be more than two) are intimately mixed.

The mechanism behind this process appears to have evolved early on, and is highly conserved: despite the effects of natural selection, it has not greatly changed over many hundreds of millions of years. This implies that it must have had been so biologically significant that any mutations in the associated DNA code were weeded out, so it is actually conserved *because of* natural selection.

At first sight this intimate mixture of fates seems puzzling, but there is a relatively easy way to achieve it: instruct each cell 'Be different from your neighbours'. This mechanism is known as lateral inhibition. An example is the nervous system. Since nerve cells form networks with long thin connections, and their ability to function depends on this geometry, it's not a good idea for the near neighbours of nerve cells to also become nerve cells. Experiments support the notion that when a cell develops into a nerve cell, it sends signals to nearby cells telling them not to do the same.

You can sometimes spot the genes responsible: if they experience a mutation, the process goes wrong and the mixed pattern fails to develop. Such observations have been recorded in the geneticists' favourite organism, the fruit fly *Drosophila*. But even if the gene responsible for lateral inhibition is known, one further puzzle remains: which cell initiates the process. In order to inhibit a neighbour, a cell must already be actively differentiating. Is the starting point also determined by some gene, or can lateral inhibition alone produce the required mixture of cell types? This is the problem that the four mathematical biologists tackled.

Experiments show that the main gene responsible for lateral inhibition is one known as *Notch*, and it acts in concert with another gene, *Delta*, that triggers the formation of neurons. Both of them generate molecular signals, in the form of proteins, that can pass from cell to cell. So the mathematical model needs to take account of these genes and how they interact and are transmitted from cell to cell. The researchers looked at two spatial arrangements of cells: a line of adjacent cells, and a plane covered in a hexagonal array, like a honeycomb (see Figure 68). Both arrangements are idealised compared with real tissues, but they capture the main features of the system in a simple way. Both are networks: place a dot at each cell and draw edges from that cell to its immediate neighbours.

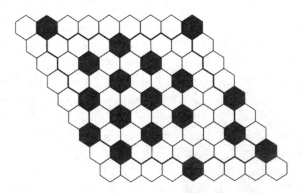

Fig 68 Pattern of primary-fate cells (black: high *Notch* activity) and secondary-fate cells (white: low *Notch* activity) in a hexagonal lattice.

The team set up suitable equations and used a computer to solve them numerically. The results confirmed that if there is strong enough feedback in the system, then initial tiny differences between neighbouring cells will automatically be amplified. In fact, this is another instance of symmetry breaking, analogous to the way slight variations in the height of a flat desert are amplified by the wind to create huge dunes. Here, the pattern is one of gene activity: cells with high levels of *Delta* activity and low levels of *Notch* activation are scattered among cells with low *Delta* activity levels and high *Notch* activation levels. These represent the two distinct fates of the original homogeneous array of cells.

The simulations often led to irregular patterns, but in every case the irregularities took the form of two adjacent cells with high levels of *Notch* activity. Two neighbouring cells with low levels of *Notch* activity never occurred. Experiments show the same effect: lateral inhibition causes primary-fate cells to be separated by at least one secondary-fate cell, but secondary-fate cells may be adjacent to one another.

The main conclusion answers the question raised earlier: it is not necessary to specify the cell that initiates the pattern. Instead, random fluctuations will amplify some initial tiny difference, producing large-scale patterns. Similarly, a desert does not have to be told which grain of sand will trigger dune formation.

16 The Paradox of the Plankton

The uppermost layers of Earth's oceans teem with plankton, organisms ranging from microscopic creatures to small jellyfish. Many are the larvae of much larger adults. They all occupy the same kind of habitat and compete for much the same resources, which is why they are all classed together under a single catch-all name. However, there is a long-standing biological principle, introduced in 1932 by the Russian biologist Georgyi Gause: the principle of competitive exclusion. This states that the number of species in any environment should be no more than the number of available 'niches' – ways to make a living. The reasoning is that if two species compete for the same niche, then natural selection implies that one of them will win.

This is the paradox of the plankton: the niches are few, yet the diversity is enormous.

The paradox is a problem in ecology: the study of systems of coexisting organisms. Although it is often convenient for biologists to study a given organism in isolation, as if nothing else existed, the real world isn't like that. Organisms are surrounded by, and often inhabited by, other organisms. The human body contains more bacteria – useful bacteria, vital to such functions as digesting food – than it does human cells. Rabbits coexist with foxes, owls and plants. These creatures interact with one another, often strongly: rabbits eat plants, while foxes and owls eat rabbits. Indirect interactions also occur: owls don't eat foxes (except perhaps very young ones), but they do eat the rabbits that the fox is hoping will make the next meal. So the presence of owls has an indirect effect on the fox population.

In 1930 the British botanist Roy Clapham recognised the interrelated nature of living creatures by coining the word 'ecosystem'. This refers to any relatively well-defined environment, plus the creatures that inhabit it. A woodland and a coral reef are both ecosystems. In a sense, the entire planet is an ecosystem: this is the essence of James Lovelock's famous Gaia hypothesis, often stated as 'the planet is an organism'. In recent years it has been recognised that if we are to ensure the continued health of the global ecosystem, and of its important subsystems, we need to understand how ecosystems work. What makes them stable, and what factors create or destroy diversity? How can we exploit the oceans without making many species of fish extinct? What effect do pesticides and herbicides have, not just on their targets, but on everything around them? And so a new branch of science was born: ecology, the study of ecosystems.

An apparently different, but closely related, branch of biology is epidemiology, the study of diseases. The subject can be traced back to Hippocrates, who noticed that there was some kind of connection between disease and environment. He introduced the terms 'endemic' and 'epidemic', to distinguish diseases that circulated within a population from those that came from outside. Modern examples in the UK are chickenpox and influenza, respectively. Epidemiology is similar to ecology because it also deals with populations of organisms within an environment. However, the organisms are now microorganisms such as viruses and bacteria, and the environment is often the human body. The two subjects start to overlap when transmission from one person to another comes into play, because now we have to consider populations of people as well as populations of disease organisms. It is, then, no great surprise to find that similar mathematical models arise in both subjects, which is one reason to treat them as variations on the same overall theme.

* * * * * * * * * * *

A basic problem in both areas is to understand how populations of organisms change over time. In many parts of the world we find boom-and-bust cycles, where a population of, say, gannets grows rapidly, exceeds the available food supply, and crashes, then repeats the same process. The resulting 'cycle' need not repeat exactly the

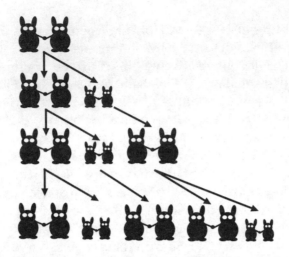

Fig 69 The first few generations in Fibonacci's rabbit model.

same numbers, but it repeats the same sequence of events. This corner of ecology is known as population dynamics.

The earliest mathematical model of the growth of a population seems to be Leonardo of Pisa's famous rabbit problem of 1202, mentioned in Chapter 4 in connection with plant numerology. Start with one pair of immature rabbits. After one season, each immature pair becomes mature, and each mature pair gives rise to one immature pair (see Figure 69). If no rabbits die, how does the population grow? Leonardo, usually known by his nickname Fibonacci (son of Bonaccio), showed that the number of pairs follow the pattern

1, 1, 2, 3, 5, 8, 13, 21, 34, 55, 89, 144, 233, 377

in which each number after the first two is the sum of the two that precede it. As we saw, these are called Fibonacci numbers. They have many interesting features; for example, the nth Fibonacci number is very close to $0.724 \times (1.618)^n$.[1] So Fibonacci's little puzzle predicts exponential growth: as we go further and further along the sequence, we find that each successive number is (very close to) the previous one multiplied by a constant amount, here 1.618.

The model is of course not realistic, and was not intended to be. It assumes that rabbits are immortal, that the rules for the birth of new rabbits are universally obeyed, and so on. Fibonacci didn't intend it to tell us anything about rabbits: it was just a cute numerical problem in his arithmetic textbook. However, modern generalisations, known as Leslie models, are more realistic: they include mortality and age structure, and have practical applications to real populations. More about these models shortly.

For large populations, it is common to employ a smoothed or continuum model, in which the population is represented as a proportion of some notional maximum population, which means that it can be thought of as a real number. For example, if the maximum population is 1,000,000 and the actual number of animals is 633,241, then the proportion is 0.633241, and the discrete nature of the population is visible only in the seventh decimal place. That is, all digits from that point on are zero, whereas in a true continuum they could take any values.

One of the simplest such models of the growth of a species of organism is the logistic equation.[2] It states in mathematical formulas that the growth rate of the population is proportional to the number of animals, subject to a cut-off as that number approaches the carrying capacity of the environment – a notional upper limit to the size of a sustainable population. The solution is known as a logistic or sigmoidal (S-shaped) curve, and it can be described by an explicit mathematical formula. The population starts near zero. At first it increases almost exponentially, but then the growth rate starts to level off. The rate of increase of the population reaches its highest value, and then starts to decrease. Eventually the size of the population levels off at a value that gets ever closer to, but never quite reaches, the carrying capacity. The maximum growth rate occurs when the population is precisely half the carrying capacity.

If a population of animals obeys the logistic equation, you can observe when the growth rate peaks, and that enables you to predict that its final size will be twice as big. Figure 70 is a classic

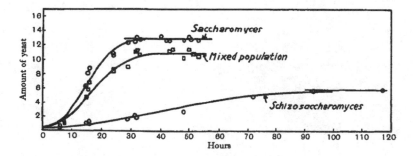

Fig 70 Gause's observations of the growth of yeast.

example, from Gause's *The Struggle for Existence*, showing growth curves for two yeast species, *Saccharomyces* and *Schizosaccharomyces*, derived from 111 experiments. It also shows the pattern of growth when the two species coexist.

. . . .

The logistic growth pattern is not realistic in many circumstances, and many other models of population growth have been devised. The principles underlying these models are relatively simple: at any given instant the total population in the immediate future must be the population now, plus the number of births, minus the number of deaths.

Leslie models, the more realistic generalisations of Fibonacci's rabbits, provide a simple example of how these principles can be implemented. They are named after Patrick Leslie, an animal ecologist who developed them in the late 1940s. They are based on a table of numbers called the Leslie matrix. A simple example captures the basic ideas, and practical models are more elaborate versions of the same thing.

Suppose we modify Fibonacci's set-up to allow three age classes of (pairs of) rabbit: immature, adult and elderly. We let time tick by in discrete steps, 1, 2, 3, and so on, and assume that at each step immature pairs become adult, adults become elderly and elderly ones die. Additionally, each adult pair gives birth, on average, to some number of immature pairs (which may be a fraction because we're working with the average). Call this the birth rate, and assume for the sake of illustration that this is 0.5. Immature pairs and elderly ones have no offspring.

The state of the population at any given time-step is given by three numbers: how many immature, adult and elderly pairs there are. Moreover, at the next time-step:

● The number of immature pairs is equal to the previous number of adult pairs multiplied by the birth rate.

● The number of adults is equal to the previous number of immature pairs.

● The number of elderly pairs is equal to the previous number of adult pairs.

These rules can be turned into a table of numbers, which in this case looks like this:

$$\begin{pmatrix} 0 & 0.5 & 0 \\ 1 & 0 & 0 \\ 0 & 1 & 0 \end{pmatrix}$$

This is called a Leslie matrix, and shows how the three age classes included in the model change at each time-step. In order from left to right and top to bottom, the age classes are immature, adult, elderly. The entry in a given row and column tells us what proportion of pairs in that column become, or give birth to, a pair whose age class corresponds to the chosen row. For example, the top row (0, 0.5, 0) says that we get 0 immature pairs from each immature pair, 0.5 immature pairs from each adult pair and 0 immature pairs from each elderly pair.

The Leslie matrix encodes the rules for all transitions among age classes, which can be more complicated than the ones I chose. There might, for instance, be ten age classes, and most of those might have various non-zero birth rates. The top row would then become a longer sequence of specific, usually different, numbers. A formula that incorporates this matrix can then be used to calculate how the numbers of pairs in the three age classes change over time.

A theoretical analysis of this formula reveals that for *any* model of this kind, there is a unique 'steady' age structure and overall growth rate, and that almost any initial choice of numbers for the various age classes ends up looking like this steady state.

For my choice of matrix, the steady age structure is approximately 23% immature, 32% adult and 45% elderly. The total population drops by 29% at each time-step (which reflects the low birth rate of 0.5, below replacement level). So this population of rabbits will eventually die out, rather than exploding like Fibonacci's rabbits.

If instead the birth rate were 1, the population would approach a fixed size; if the birth rate were larger than 1, the population would explode. This tidy transition at birth rate 1 arises because only adults have offspring. With more age classes there are several birth rates, and the change from dying out to exploding is more complicated.

. . . .

An important application of such models is the growth of the human population, currently estimated to be just under 7 billion. Very sophisticated models are needed to predict future growth, because this depends on age distribution, social changes, immigration, and many other social and political features. But all models must obey the basic 'law of conservation of people': people can be created by birth, they can be destroyed by death and they can move from one nation to another, but they can't (astronauts excepted) vanish into thin air.

This law can easily be turned into mathematical equations, but the form of the equations depends on the birth rate and the death rate, and on how these change as the population itself changes. Leslie models use constant birth rates for each age class, and split the population into a fixed number of age classes. Other models replace these assumptions by more realistic ones: for instance, the birth rates might depend on the overall population size, or people might remain in a given age class for a certain time before moving to the next one.

A good model requires realistic formulas for all birth and death rates. These can be obtained if good data are available, but for the world population accurate data exist only from 1950 to the present. This is too short a time span to determine, with any certainty, the specific form that the equations should take. So experts make informed guesses and choose what seems most reasonable. Not surprisingly, different experts prefer different models. Some use deterministic models, some use statistical ones. Some combine the two. Some are orthodox, some not.

In consequence, there is a lot of disagreement about when the Earth's population will peak, and how big it will be when it does. Predictions range from 7.5 billion to 14 billion. Figure 71 shows the growth since 1750, with a short prediction in the middle of that range. The evidence suggests that the rate of growth has been fairly constant since the 1970s, when the population reached 4 billion, so there is no clear evidence that the growth rate is slowing down, let alone that it will level off or start to decline. Nonetheless, the world population is generally expected to peak some time in the next 150 years. The main reasons for this expectation are social and cultural. An important one is the 'demographic transition', in which improved education and standard of living cause a sharp fall in the

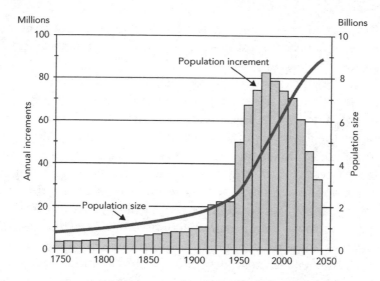

Fig 71 Long-term world population growth, 1750–2050 (predicted after 2010.) The curve shows population size, bars show increases in size at intervals of 2.5 years.

size of families. But in many countries this is offset by rising life expectancy, caused by improvements in medicine and standard of living.

It is difficult to incorporate these effects into population models, because they depend on scientific advances, political changes, and cultural shifts, all of which are inherently unpredictable. Statistical methods are often used, and like all statistics they work better for large populations. So, just as it turns out to be easier to forecast the global climate than to predict local weather, it is easier to forecast the general trend of the global population than it is to predict national populations. But even then, the uncertainties are huge.

· · · · · · · · · · · ·

The traditional states seen in dynamical systems are steady states (also called equilibria), where nothing changes as time passes, and periodic states, where the same sequence of events repeats over and over again. A rock is in a steady state if it's not moving and we ignore erosion. The cycle of the seasons is periodic, with period one year. But in the 1960s mathematicians realised that tradition had completely missed another, more puzzling kind of behaviour: chaos. This is behaviour so irregular that it may appear random, but it arises in models without explicit random features, for which the present completely determines the future. Such models include all dynamical systems.

At first, many scientists viewed chaos with suspicion, presumably because they thought that such outlandish behaviour had no place in nature. But chaos is entirely natural: it arises whenever the dynamics of a system mixes it up, much like kneading dough mixes the ingredients. One consequence is the 'butterfly effect', which originally arose in weather forecasting. In principle, and in a very specific sense, the flap of a butterfly's wing can change the global weather pattern. More prosaically, although the future of the system is completely determined by its present, this requires knowing the present to infinite accuracy. In practice, tiny measurement errors in determining the present state grow rapidly, making the future unpredictable beyond some 'prediction horizon'.

As soon as mathematicians started to think about dynamics geometrically, chaos became obvious. It only seems outlandish if you are looking for solutions that can be expressed by neat, tidy formulas. And those are rare.

In fact, it was geometric thinking – about the stability of the Solar System – that led the French mathematician Henri Poincaré to discover chaos in 1895. A few sporadic developments occurred during the first half of the twentieth century, but it all came together in the 1960s when Stephen Smale and Vladimir Arnold developed a systematic topological approach to dynamics. In 1975, in a survey article in *Nature*, Robert May brought these new discoveries to the attention of the scientific community, and ecologists in particular.[3] His main message was that complex dynamics can arise in very simple models of population growth. Simple causes can have complex effects; conversely, complex effects need not have complex causes.

His main example, selected for its simplicity as an introduction to these phenomena, was a variant of the logistic model in which time ticks by in discrete amounts: 1, 2, 3, and so on. This assumption is natural when studying successive generations of a population, rather than its evolution moment by moment. 'One of the simplest systems an ecologist can study,' May wrote,

> is a seasonally breeding population in which generations do not overlap. Many natural populations, particularly among

temperate zone insects (including many important crop and orchard pests), are of this kind ... The theoretician seeks to understand how the magnitude of the population in generation $t+1$, X_{t+1}, is related to the magnitude of the population in the preceding generation t, X_t.

As a specific example he cites the equation

$$X_{t+1} = X_t(a - bX_t)$$

The initial population is specified as X_0, and the formula is then used to deduce the values of X_1, X_2, X_3, and so on by successively setting t to be 0, 1, 2, 3, ... Here a and b are parameters – adjustable constants whose values may change the dynamics. For example, when b is zero the equation describes exponential growth – very similar to Fibonacci's rabbit model, except that now there is only one generation. But when b becomes larger, the population growth is restricted, modelling limitations on resources. A mathematical trick[4] simplifies the equation to

$$X_{t+1} = aX_t(1 - X_t)$$

with a single parameter a, and this is the form normally studied by mathematicians.

The behaviour of this equation depends on the parameter a, which has to lie between 0 and 4 to keep X_t between 0 and 1. When a is small, the population converges to a steady state. As a increases, oscillations set in, initially cycling through two distinct values, then 4, then 8, then 16, and so on. At a=3.8495, the regular oscillations stop and the system behaves chaotically. Chaos then predominates, but there are small ranges of a that lead to regular behaviour (see Figure 72).

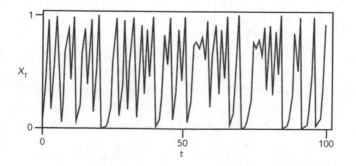

Fig 72   Chaotic behaviour in the discrete logistic model when a=4.

Although this model is too simple to be realistic, there is no reason why more complex models should not behave in similar ways, and there is plenty of evidence that they often do. So erratic changes in natural populations, previously attributed to irregularities in the environment such as changes in climatic conditions, can in fact be generated by the free-running dynamics of the population itself. May ended his paper with a call for such examples to be widely taught in schools, to prevent people assuming that irregular effects necessarily have irregular causes.

. . . .

All well and good, but do real populations exhibit chaos? In the wild, it is difficult to separate a population's own dynamics from the variations in environment that always occur in nature, so the occurrence of chaos has been controversial. Most of the data that zoologists and entomologists have collected over the years on animal and insect populations are seldom extensive enough to distinguish chaos reliably from randomness. The physical sciences get round such problems by performing controlled experiments in the laboratory, but even in the lab it is difficult to control the large number of extraneous variables that might affect experimental results in ecology. However, it's not impossible.

In 1995 James Cushing and colleagues at the University of Arizona began a series of experiments that demonstrate the occurrence of chaos in populations of *Tribolium castaneum*, known as the flour beetle or bran bug because it often infests stocks of milled grain. Their theoretical model has three variables: the number of feeding larvae, the number of non-feeding larvae (plus pupae, and newly emerged adults) and the number of mature adults.[5] The flour beetle and its larvae indulge in egg cannibalism: they eat the eggs of beetles in the same species, including their own. This behaviour is incorporated into the equations of the model.

Some of the experiments were performed under specially controlled conditions: beetles were removed or added to the population to mimic the mortality rates observed in the wild. Other experiments were not manipulated in this manner. To avoid genetic changes, the adult population was replenished from time to time from other cultures, maintained under strict laboratory conditions.

These experiments showed the expected onset of oscillations, but not chaos. However, the theoretical model can be chaotic, and the experimental system was quite close to the range of variables in which chaos occurs in the model. By changing the protocol to mimic a higher mortality rate than there would normally be in the wild, the same team managed to drive the beetle population into chaotic fluctuations.[6] This second paper concludes:

> The experimental confirmation of nonlinear phenomena in the dynamics of the laboratory beetle lends credence to the hypothesis that fluctuations in natural populations might often be complex, low-dimensional dynamics produced by nonlinear feedbacks. In our study, complex dynamics were obtained by 'harvesting' beetles to manipulate rates of adult mortality and recruitment. For applied ecology, the experiment suggests adopting a cautious approach to the management or control of natural populations, based on sound scientific understanding. In a poorly understood dynamical population system, human intervention ... could lead to unexpected and undesired results.

. . . .

Chaos also solves the paradox of the plankton. The paradox is a violation of Gause's principle of competitive exclusion: there are many more species of plankton than there are environmental niches. The plankton can't be wrong, so there must be more to the principle than is usually thought. The question is, what?

There are well-established ecological models that agree with Gause's principle. Ultimately, they trace the relation between the number of species and the number of niches to a general mathematical fact: if you have more equations to solve than you have variables, solutions don't exist. Roughly speaking, each equation pins down a relation between the variables. Once you have as many relations as variables, you can find the solutions. Any extra equation is likely to contradict the solutions already found. As a simple case, the equations $x+y=3$, $x+2y=5$ are satisfied only when $x=1$, $y=2$. If we add a further equation, such as $2x+y=3$, it is not valid for that solution. Only when the extra equation provides no new information will the original solution survive.

The competitive exclusion principle often works well, and since the mathematics supports it, ecologists have a puzzle on their hands.

Part of the answer is that the upper reaches of the Earth's oceans are vast, and plankton are not uniformly mixed. But it now looks as though there could be a better explanation of the paradox of the plankton. The standard mathematical model makes a restrictive assumption: it seeks steady-state solutions to the relevant equations. The populations of organisms in the species concerned are assumed to remain constant over time; they can't fluctuate.

This assumption in effect takes the 'balance of nature' metaphor for an ecosystem too seriously. Real ecosystems, if they are to survive for long, have to be stable. If the populations of various organisms fluctuate wildly, some may die out, and that changes the dynamics of the ecosystem. However, stability need not require the entire system to remain in exactly the same state for ever, just as a stable economy is not one in which everyone always has exactly the same amount of money as they did yesterday. The crucial feature of stability is that fluctuations in populations must remain within fairly tight limits.

Chaotic dynamics does precisely that. It exhibits erratic fluctuations, but the size and type of those fluctuations is determined by an attractor: a specific collection of states to which the system is confined. It can move around inside the attractor, but it can't escape. In 1999, the Dutch biologists Jef Huisman and Franz Weissing showed that a dynamic version of the standard model of resource competition can produce regular oscillations and chaos if species are competing for three or more resources.[7] In other words, as soon as the system is allowed to be out of equilibrium, the same resources can permit a much greater amount of diversity among the organisms that are using them. Roughly speaking, the dynamic fluctuations allow different species to utilise the same resources at different times. So they avoid direct competition not by one of them winning and killing off all the others, but by taking turns to access the same resource.

The same researchers, and others, have since developed these ideas into a wide range of models, and the resulting predictions are often in agreement with data from real plankton communities. In 2008, Huisman's team reported an experimental study of a food web isolated from a natural one in the Baltic Sea, involving bacteria, plant plankton, and both herbivorous and predatory

animal plankton.[8] Their observations were carried out over a period
of six years in a laboratory. The external conditions were kept
exactly the same throughout, but the populations of the species
concerned fluctuated significantly, often by a factor of 100 or more.
Standard techniques for detecting dynamical chaos revealed its
characteristic signs. There was even a butterfly effect: the future of
the system remained predictable only a few weeks or a month
ahead.

Their report remarks that 'Stability is not required for the
persistence of complex food webs, and that the long-term
prediction of species abundances can be fundamentally impossible.'
And it refers back to May's original suggestion that chaos could well
be important for our understanding of ecosystems – a prescient
insight that is now thoroughly vindicated.

. . . .

Disease epidemics take place in a special type of ecosystem,
involving both the organisms that become infected and the
microorganisms – viruses, bacteria, parasites – that cause the
disease. So similar modelling techniques, suitably modified, can be
used for both ecosystems and epidemics.

In 2001 an abattoir in Essex reported that a consignment of pigs
was suffering from foot-and-mouth disease. The disease spread
rapidly, and the European Union slapped an immediate ban on the
export of all British livestock. In all, there were two thousand
outbreaks of the disease on British farms, leading to the culling of
ten million sheep and cattle. The total cost was around £8 billion,
and the news media showed piles of dead cattle being burnt in
fields, a scene straight out of Dante's *Inferno* that did little for public
confidence. Was the strategy of stopping all animal movement
within the UK, and slaughtering all animals on any infected farm,
the right one?

Foot-and-mouth disease is caused by (several forms of) a virus
that hardly ever affects humans, a picornavirus (see Figure 27,
p. 141). But food is a very sensitive issue, so it would not be
acceptable to let the disease spread. It also causes damage to meat
and milk production, causes serious distress to the animals, and
leads to import bans. So the standard response throughout the
world is to eradicate it. Vaccination might become a viable and

cheaper alternative. But even if the response is to slaughter infected animals, or those that soon might be, many different strategies for controlling the disease can be contemplated.

It is therefore important to decide which strategy is best. Mathematical modelling of the 2001 outbreak, after the event, suggests that initially the UK Government's response was too slow; then, when the disease became widespread, it was too extreme. Only one animal in five among those slaughtered was infected. This overkill may have resulted from inadequate and outdated mathematical models for the spread of the epidemic.

Models are the only way to predict the likely spread of an epidemic and to compare possible control strategies. They can't forecast which farms will be hit, but they can provide an overview of general trends, such as the rate at which the disease is likely to spread. In the 2001 outbreak, three different models were used.[9] When the outbreak began, the main model available to DEFRA, the government department then responsible for agriculture, was a probabilistic one called InterSpread. This provides a very detailed model, farm by farm if need be, and includes many different routes for disease transmission. It might seem that the more realistic a model is, the better it will perform, but ironically InterSpread's complexity is also its weakness. The calculations take a long time, even with powerful computers. And fitting the model to real data requires setting the values of a large number of parameters, so the model may be unduly sensitive to small errors in estimating these parameters.

A second model, the Cambridge–Edinburgh model, can also represent the location of every farm, but it uses a much simpler mechanism to model the transmission of the disease. Farms with the disease are 'infectious', those that are not yet infected but may come into contact with infected farms are 'susceptible', and the model combines all these variables to come up with an overall measure of how rapidly any given farm is likely to infect others. This model forecasts the geographical spread of the disease quite well, but its performance is poorer when it comes to the timing – perhaps because it assumes that the disease takes the same time to show up in all infected animals, and every animal remains infectious for the same period of time. In reality, these times vary from one animal to another.

The third model, the Imperial model, is based on traditional equations for the spread of epidemics, and was put together during the epidemic. It was less realistic than the other two, but much faster to compute, so it was more suitable for tracking the progress of the disease in real time. It predicted the changes in the number of infected animals, but not the locations of outbreaks.

Each model turned out to be useful for some types of forecast, and subsequent analysis suggests that, on the whole, the strategy of widespread slaughter was probably correct. It would not have been feasible to determine precisely which animals were infected during the rapid spread of the disease, and any infected animals mistakenly left alive would create new centres from which the disease could again spread, rendering previous actions useless. But the analysis also made it clear that the initial response was too slow. If more stringent restrictions on the movement of animals had been put in place immediately, and early cases spotted sooner, then the disease would not have had such a huge economic impact.

The second and third models also indicate that vaccination is not likely to be an effective control strategy once the disease becomes widespread, but vaccinating all animals in a ring surrounding an infected farm might restrict the spread of the disease if it is done right at the start.

These three models of the foot-and-mouth epidemic show how mathematics can help to answer biological questions. Each model was much simpler than any truly 'realistic' scenario. The models did not always agree with one another, and each did better than the others in appropriate circumstances, so a simple-minded verdict on their performance would be that all of them were wrong.

However, the more realistic the model was, the longer it took to extract anything useful from real-world data. Since time was of the essence, crude models that gave useful information quickly were of greater practical utility than more refined models. Even in the physical sciences, models mimic reality; they never represent it exactly. Neither relativity nor quantum mechanics captures the universe precisely, even though these are the two most successful physical theories ever. It is pointless to expect a model of a biological system to do better. What matters is whether the model provides useful insight and information, and if so, in which circumstances. Several different models, each with its own strengths and weaknesses, each performing better in its own particular context, each providing a significant part of an overall picture, can be superior to a more exact representation of reality that is so complicated to analyse that the results aren't available when they're needed.

The complexity of biological systems, often presented as an insuperable obstacle to any mathematical analysis, actually represents a major opportunity. Mathematics, properly used, can make complex problems simpler. But it does so by focusing on essentials, not by faithfully reproducing every facet of the real world.

17 What is Life?

. . . .

Biology is the study of life, in all its forms – on this planet.

Since we currently know of no other place in the universe where life definitely exists, or used to exist, 'on this planet' may seem superfluous. However, it points to a gap in today's biological knowledge, one that would be present even if there were no life anywhere else in the universe.

In its strongest form, the gap is the general question 'What is life?' Must all life *in principle* be similar to the living creatures of this planet – built from carbon chemistry, controlled by DNA, composed of cells ... in short, just like us? Is there no alternative, not even hypothetically? Or could entities that reproduce, and are sufficiently complex and organised to qualify as 'living', be made from different materials, be organised in different ways? More strongly, do such entities exist, somewhere in our galaxy or another galaxy?

More bluntly: could aliens exist – and do they?

The first part of this question is a lot easier than the second. We can investigate the potential for exotic life forms without exploring the planets that circle distant stars. But even then, we run into deep problems. We've already seen that biologists disagree about whether viruses are alive, so 'life' is to some extent a matter of definition. To answer the second part, we have to make contact with extraterrestrial life – whatever we decide that means. We might do this by visiting another world, by observing chemical signatures of living processes through powerful telescopes, by receiving messages from an alien civilisation, or by waiting for aliens to visit us.

In the next chapter I will argue that UFO reports and claims of alien abduction are not sufficiently convincing to conclude that the fourth of those options has already occurred. The SETI project has been pursuing the third option since 1961,[1] so far without success, but it could pay off at any moment if advanced aliens actually do exist. We are just beginning to attempt the second option. The first is now being done using robotic explorers, and is currently confined to our own Solar System. Humans last landed on the Moon in 1970, and the proposed project to land a human on Mars has been cancelled.

. . . .

Some biologists define life to be just like Earthly life: based on carbon, water, organic chemistry, DNA, proteins, the whole shebang. It works here, but this surely begs the point, by making a gigantic assumption for which no evidence exists. Worse, such a concept of life has to be continually adjusted as we discover more about our own planet's more exotic inhabitants. A lot of Earthly life is distinctly different from what we believed was 'normal' fifty years ago.

I can't help imagining two cavemen discussing the definition of 'tool'. They quickly agree on two fundamental points: a tool has to be made of flint and it has to fit in your hand. Otherwise there would be no way to make it, and people would not be able to use it. Now imagine their faces if some time-traveller turns up with a bulldozer.

If we are to discuss potential forms of life – like ours or not – the first step is to agree on a working definition of 'life', and I'll spend the rest of this chapter on that question, returning to aliens in the next one. Life is one of those annoying concepts that can usually be recognised when you see it, but turns out to be hard to pin down precisely. I don't personally find that to be either a surprise or an obstacle: in my experience, the only scientific concepts that can be pinned down with absolute precision are in areas that were mined out long ago. Think of all the fuss about whether Pluto counts as a 'planet'. Even in mathematics, where precise definitions are de rigueur, it is common for them to evolve as new research reveals new aspects. We've already seen this for such basic terms as 'space' and 'dimension'.

Biologists don't have a universally accepted definition of 'life';

instead, they have several competing definitions, none totally satisfactory. At one extreme, you can build carbon chemistry and DNA explicitly into the definition. If you do, then anything acceptable as a life form will use carbon chemistry and DNA – end of story. However, this begs most of the interesting questions. From the viewpoint of mathematics and physics, terrestrial life looks like an example (rather, a gigantic number of closely related examples) of what ought to be a far more general *process*. Many biologists feel the same way: they dislike defining life in terms of what it is made of: what it does and how it works seem more appropriate and less limiting. It's rather like having a mathematics that is limited to numbers between 1 and 100, and wondering whether a more general concept of number could preserve most of the interesting properties observed in that range.

The upshot is that the current working definitions of life concentrate on what it does, rather than what it is. The main features of life are:

- possessing an organised structure;
- regulating internal behaviour in response to short-term changes in the environment;
- maintaining both the above by extracting energy from the environment;
- responding to external stimuli, say by moving towards a food source;
- growing – in a way that does not merely accumulate more and more stuff while doing nothing with it;
- reproducing;
- adapting to long-term changes in the environment.

These are not the only things that living creatures do, they are not mutually exclusive, some are less important than others, and some might even be dispensed with altogether. But in broad-brush terms, if some system in nature exhibits most of the features on the list, then it may qualify as a form of life.

To appreciate the difficulties, think about a flame. Flames have a definite physical structure. They change their dynamics in response to their surroundings, growing in the presence of fuel and oxygen, dying down if these are absent. They extract chemical energy from the reaction between fuel and oxygen. They invade adjacent sources of fuel. They grow. They reproduce: a forest fire starts as a single flame. But the chemistry of flames today is the same as it was a billion years ago, so they fall at the final hurdle.

With a vivid imagination, you could invent plausible aliens that were complex systems of flames. If their chemistry could change over long periods of time, depending on what the environment can provide, they might evolve. In a way, we are like that: the energy cycle that keeps our cells working – and us – is an internalised flame. It is an exothermic reaction: it gives off heat. But we are more than just an exothermic reaction.

Many alternatives to this list of properties of life have been proposed. Nearly all of them feature the most obvious of life's characteristics: reproduction. They distinguish this from a closely related ability: replication. The distinction is important – indeed, vital.

An object or system *replicates* if it makes exact copies of itself, or copies that are so similar that it is hard to tell the difference. It *reproduces* if the copies have some degree of variability. A photocopier replicates black-and-white text documents: aside from differences in paper, and possible enlargement or reduction, the copy is essentially the same as the original. In particular, the text, which is what usually matters, is essentially identical, though even here there are smudges and gaps: true replication in the strongest sense is rare. Even copies of computer files may contain errors. In contrast, a cat reproduces: its kittens, even when fully grown, do not resemble it in detail, and they often have totally different markings, sizes and gender. But they grow into cats and can, on the whole, father or give birth to their own kittens. So the *system* 'cat' reproduces, but does not replicate.

One of the pithiest and potentially broadest definitions of life was devised by Stuart Kauffman, and it is based on thinking of a living organism as a complex system. This phrase has a specific technical meaning: something composed of a large number of relatively simple agents, or entities, which interact according to relatively simple rules. The mathematics of complex systems shows that despite the simplicity of the ingredients, the combined system often (indeed usually) displays complicated 'emergent' behaviour, not evident in the entities and rules. In Kauffman's view, life is a complex system that can reproduce, and can carry out at least one thermodynamic work cycle (which converts heat into work and can be repeated over and over again, given a reliable heat source). All other properties of life can be seen as possible – sometimes inevitable – consequences of these basic features. This definition emphasises general mathematical features of the system, not the ingredients it is made from. However, it has yet to find favour among the majority of biologists.

. . . .

It is of course fascinating to realise that life on Earth is heavily dependent on the information-bearing properties of DNA, and only a scientist who is brain-dead could fail to want to know how it all works. But this does not imply that when the question is 'Life?', the answer is 'DNA'. If the question were 'Cake?', few scientists would settle for the answer 'Baking soda'. They'd want to know just what the baking soda contributed, and whether anything else might have a similar effect. So it can't be the details of terrestrial biochemistry that really explain why life is possible: it must be the abstract process that the biochemistry realises.

Life on Earth is one example, the only one we know. The question is: an example of what?

Must life be based on DNA? If you think that DNA is *the* key to life, you *have* to answer 'yes'. Numerous interesting questions crop up, even then. Would the 'genetic code' that turns DNA triplets into proteins have to be the same as the code used on Earth? The code is implemented by transfer RNA, and experiments show that transfer RNA could implement a different code with the same facility, using the same chemical processes. You can synthesise non-standard transfer RNA and the whole system still functions perfectly well. So let us accept as a working hypothesis that in principle Earthly life does not employ the only system that would work.

Upping the ante slightly, it is also clear that any molecule sufficiently similar to DNA could play a similar role. There could be minor variants in which different bases occurred: even on Earth some viruses employ RNA in place of DNA, and RNA uses uracil instead of thymine; RNA also plays several key roles in the reproduction of nearly all organisms. Synthetic 'exotic bases' have been created in the laboratory and inserted into DNA double helices. The code has been extended to a four-base code, in the laboratory, so the range of amino acids encoded could be in the hundreds. So it seems unlikely that DNA is the only possible linear 'information-bearing' molecule (more properly, class of molecules, since the whole point is that different organisms have different DNA).

Nor need an information-bearing molecule be linear – tree-like structures, or two- or three-dimensional polymer arrays, could also be used. Because of the unique properties of carbon, it seems probable that any large information-bearing molecule would have to be organic (carbon-based) – but silicon could be a possible alternative in the presence of the occasional metal atom, which stabilises complex silicon-based molecules. In fact, the kind of organised complexity that is required to make a living creature could in principle be found in, say, magnetohydrodynamic vortices in the photosphere of a star, crystalline monolayers on the surface of a neutron star, wave-packets of electromagnetic radiation crossing the wastes of interstellar space, quantum wave-functions, or even non-material creatures inhabiting pocket universes whose physics is very different from ours.

Whether or not such creatures actually exist, the thought experiment 'What if they did?' suggests that there might be a meaningful definition of 'life' that is valid in much greater generality than 'organised chemical systems that reproduce using DNA'. This would be a definition that emphasised the abstract processes of life, not special material constituents. Such a definition would presumably involve the apparent ability of living creatures to self-organise and self-complicate.

. . . .

Until the middle of the twentieth century, the list of properties used to define life could have been shortened to just one: reproduction. (Some minor tweaks in what that word means would rule out flames.) No known non-living object, or system, could replicate, let alone reproduce. So living systems had a unique, mysterious ability which non-living systems could not emulate. Early proposals for the cause of this ability included possession of a soul or the presence of some vital force (*élan vital*) that animated non-living matter. But for scientists, explanations like these are unsatisfactory unless the soul can be located somewhere in the organism, or the vital force can be identified. Otherwise we are back to the medieval habit of explaining why objects fall by appealing to some tendency to 'move to their natural place' – on the ground. They fall because they move towards the ground ... brilliant!

By the middle of the twentieth century it had become clear that the ability to replicate is a system property: an ability arising from the way the system is organised. Replication is a consequence of the system's structure; it does not require some magic extra ingredient that no one can locate or identify. The system that proved this definitely had no such ingredient, because it was a mathematical abstraction. Its inventor was John Von Neumann, whom we have already encountered as the creator of game theory.

Von Neumann's first step in this direction came in lectures he gave in 1948, and it was a thought experiment. Imagine a programmable robot living in a warehouse filled with spare parts, which it can manipulate. It also has a tape containing instructions. The instructions tell it to wander round the warehouse and build a copy of itself by picking up all the parts required – except for the tape. Finally, it makes a duplicate of the tape and inserts this into the copy.

There are problems with this description: for example, the robot has to be able to copy the tape, so isn't replication built in from the start? Von Neumann's point was that in this scenario, neither the robot alone, nor its program, can replicate. Only the combined system does that. The program replicates the robot; the robot replicates the program. This division of roles was a key insight, because it got round what had previously seemed to be an insurmountable logical obstacle to a self-replicating device.

Suppose that a self-replicating entity exists. Then it must contain within itself a complete specification of its own structure, in order to know how to construct the copy. However, the copy must also be able to replicate (or else it's not a true copy), so this internal specification of structure must contain something that specifies the part of the copy that provides a complete specification of its own structure ...

More carefully: the device must contain a *representation* of itself. Inside that representation must be a representation of the second-generation device. Inside that must be a representation of the third-generation device, and so on. So it looks as though any self-replicating device has to resemble a set of Russian dolls, each nested inside the previous one, and this set of dolls must go on *for ever*. If it didn't, the smallest doll would not contain a representation of itself, and so could not replicate.

No physical entity can do this: at some point, the doll has to become smaller than the smallest fundamental particle. Of course no living entity can do it either, but that's not a problem if you believe that life rests on some supernatural 'essence'. The supernatural doesn't have to make sense.

Von Neumann's proposal for the architecture of his device avoided the Russian-doll objection. It did so by interpreting the same physical object – the program on the tape – in two conceptually distinct ways. In one, the program consisted of instructions, to be obeyed. In the other, it consisted of symbols, to be copied. In one, a piece of paper with 'put the kettle on' causes a kettle to be boiled. In the other, it leads to a second piece of paper bearing the message 'put the kettle on'.

Now the program can replicate the robot when the robot obeys the instructions, and the robot can replicate the program by copying it but *not* obeying the instructions.

. . . .

Von Neumann wasn't satisfied with this set-up, because he couldn't see a good way to analyse it mathematically, or to build a real machine that could carry it out. At the time, he was working at the Los Alamos National Laboratory in New Mexico, and one of his colleagues was the mathematician Stanislaw Ulam. Ulam, renowned for his original turn of mind, had been modelling the growth of crystals using a lattice: a square grid like a large chessboard, without the chequered pattern. He suggested that Von Neumann might be able to implement his self-replicating machine by employing a similar trick. In detail, Ulam's idea was to define a self-replicating cellular automaton (see Figure 73).

In this context, an automaton is a mathematical system that can obey simple rules – in effect, perform elementary computations. A cellular automaton is a grid with rules, like a simple video game. It is a special kind of complex system, with cells as entities and – well, rules as rules. Each square on the grid – each cell – can exist in a variety of states. One way to visualise the states is to colour the cells, so that possible states correspond to a list of colours. Each cell

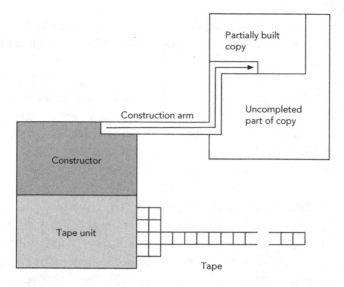

Fig 73 Schematic of Von Neumann's replicating automaton.

now obeys a specific system of rules, in which its own colour and those of its neighbours determines the next colour for that cell. For instance, with two colours 'red' and 'blue' the rules might be a list of statements like this:

- If you are red and your four immediate neighbours are all blue, turn blue.
- If you are red and exactly three of your neighbours are blue, turn blue.
- If you are red and exactly two of your neighbours are blue, remain red.

The full list would cover all possible patterns of states.

With the colours and rules in place, you start the automaton in some pattern of colours (initial state), apply the rules (simultaneously on all cells) to get the next pattern, repeat to get the pattern after that, and so on. It sounds simple, but the consequences can be complex. A suitable cellular automaton can mimic any calculation that a real computer can carry out.

Inspired by Ulam's suggestion, Von Neumann worked out a set of rules for an automaton with 29 cell colours.[2] The replicator device occupied about 200,000 squares; the rest were left blank – in effect, another colour, which changed only if a neighbouring cell ceased to be blank. Von Neumann proved that by following the

simple list of rules, the automaton would build a copy of itself.
Which would then build another copy, which would build another
...

He never published his results – he may have seen them as a
diversion from his main research, or he may have lacked the time
or the inclination. Whatever his reasons were, it's a pity that he
didn't put his ideas into print, because they would have constituted
an important mathematical prediction about real organisms:
namely, when an organism reproduces, it must employ some list of
data (the tape) that has two distinct functions – to control the
replication procedure, and to be copied. The discovery of the
structure of DNA, and its role in the reproduction of organisms,
would have verified that prediction. As it happens, Von Neumann's
work came to public attention only in 1955, just after Crick and
Watson's epic paper. And it was not until 1960 that the American
mathematician and computer pioneer Arthur Burks gave the first
complete proof that Von Neumann's mathematical machine could
replicate.[3] So the chance to predict a basic mechanism of biological
reproduction from general mathematical principles went begging.

• • • • • • • • • • • •

Several people took up Von Neumann's ideas. Conway (whom we
last came across in knot theory) was among them: he invented a
cellular automaton with dynamics so flexible and 'unpredictable'
that he named it the Game of Life.

'Life', as it is usually known, is played with counters on a square
grid. The game begins by setting up some finite configuration of
counters, the initial state of the automaton. Then a short set of
simple rules, involving the number of immediate neighbours of
each counter, is applied to get the next configuration. These rules
govern the survival, birth or death of counters. Dead counters are
removed from the grid, newborn ones are added and the rest stay
where they are.

The precise rules are:

- A counter with 0 or 1 neighbours dies.
- A counter with more than 3 neighbours dies.
- A counter with 2 or 3 neighbours remains alive.

- An empty space with exactly 3 neighbouring counters gives birth to a new counter.

There is a host of information about Life on the Internet, together with free software to run the game.[4] Life runs on rigid rules, so the future of any given initial configuration is completely determined: if the game is run again starting from the same shape, the subsequent history is the same as before. Nevertheless, the outcome is unpredictable in the sense that there is no short cut that can predict what happens – all you can do is run the game and see what evolves. This is one of several ways in which 'deterministic' and 'predictable' differ in practice, despite being essentially the same in principle.

Despite the simplicity of the rules of Life, its behaviour can be astonishingly rich. So rich, in fact, that it is sometimes unpredictable in a very strong sense, even though the initial state completely determines everything that happens later. In 1936, Alan Turing provided a solution to the halting problem, proving that in general it is not possible to forecast ahead of time whether a computer program will terminate with an answer, or go on for ever – for example, getting stuck in a loop and repeating it indefinitely. Conway and others proved that there exists a configuration in Life that forms a universal Turing machine, a mathematical representation of a programmable computer.[5] So there is no way to predict whether a given Life configuration will live for ever or die out.

In 2000, Matthew Cook found a simpler universal Turing machine, by proving a conjecture that the English polymath Stephen Wolfram, had made in 1985: a cellular automaton whose states form a line of cells, rather than square grid, can also mimic a universal Turing machine.[6] This automaton is known as 'Rule 110'. It has two states, say 0 and 1, and its rules are very simple. To find the next state of a cell, look at that cell and its two neighbours to the left and the right. If the pattern is 111, 100 or 000, that state becomes 0; otherwise, it becomes 1. It is remarkable that such a simple system of rules can in principle do anything that a computer can do – for example, calculate π to a billion decimal places. This reinforces the main message of artificial life: never underestimate the complexity of the behaviour that can result from simple rules.

• • • • • • • • • • • •

When Conway invented the Game of Life, Chris Langton, who at the time was working in a hospital programming mainframe computers, found the game so interesting that he began to experiment with computer simulations of features of living creatures. Burks was running a postgraduate programme at the University of Michigan, and in 1982 Langton joined it. The outcome was a new sub-branch of science: artificial life. Some people object that this name is an exaggeration, but it should be obvious that the name is not intended to indicate the creation, by artificial means, of *real* life. Instead, it refers to non-biological systems that mimic, or emulate, some of the key features of living organisms, such as replication. Those features are puzzling in their own right, independently of their physical realisation, so it makes sense to study them in mathematical systems that separate the features from what they are made from.

Langton described the new field at the first conference on the topic, saying

> Artificial life is the study of artificial systems that exhibit behavior characteristic of natural living systems. It is the quest to explain life in any of its possible manifestations, without restriction to the particular examples that have evolved on earth ... The ultimate goal is to extract the logical form of living systems.[7]

As a demonstration of what was possible, Langton had already invented the first self-replicating 'organism' to be implemented in a real computer. But replication is only one of the puzzling features of living organisms. Reproduction – replication with occasional errors – opens up the possibility of evolution; all that is required is a selection principle, to decide which changes to keep and which to discard.

Over the past thirty years, a seemingly endless stream of artificial life systems, defined in many different ways, has made three things abundantly clear – all of them contrary to most previous intuition:

1. Almost any rule-based system, capable of any kind of behaviour more complex than steady states or periodic cycles, is capable of very complex behaviour indeed. In rule-based systems, complex behaviour is the norm.

2. There is no significant connection between the complexity or
    simplicity of the rules, and the complexity or simplicity of the
    resulting behaviour. Complex rules can lead to behaviour that is
    simple, or complex. Simple rules can lead to behaviour that is
    simple, or complex. There is no 'conservation of complexity'
    between rules and behaviour.

3. Evolution is a remarkably powerful way to create highly
    complex structures and processes, without designing the desired
    features into the evolving entities in any explicit way.

Langton's basic idea has been implemented in many different
forms, in systems with names like Tierra, Avida and Evolve.
Darwinbots, introduced in 2003 by Carlo Cormis, is typical.[8]
Individual organisms, the aforementioned bots, are represented on
the computer screen as circles. Each bot is equipped with simulated
genes, which affect its behaviour. It acquires energy by feeding, and
its energy runs down as it carries out its activities. If the energy
level gets too low, it dies. The bots display a rich variety of
behaviour. Unlike the cells in Life and Rule 110, they can wander
around all over a plane, and are not confined to specific, discrete
cells.

The philosophical stance known as 'weak alife' holds that the
only way to create a living process is through chemistry. ('Alife' is
jargon for artificial life.) Since we already know that the standard
DNA-based system that occurs naturally on Earth can be changed
and will still work, it is not possible to retreat further and insist that
the form of life that we know is the only kind that is possible.
However, we have no solid evidence of non-chemical life, so it is
reasonable to argue that all life must be chemical.

The main message of artificial life is more imaginative and more
speculative. It goes back to Von Neumann, and is called 'strong
alife'. This position maintains that life is not a specific chemical
process, but a general *type* of process that does not depend on the
medium used to implement it.

If strong alife is right, what matters is not what life is made
from, but what it does.

. . . .

Synthetic life may sound rather similar to artificial life, but the term refers to organisms that run on conventional biochemistry but are synthesised in the laboratory from inorganic ingredients.

In 2010 a team at the J. Craig Venter Institute in Rockville, Maryland, announced the creation of an organism nicknamed Synthia. First, the team made a copy of the genome of the bacterium *Mycoplasma mycoides*, 1.2 million base pairs, by purely chemical techniques – no living organisms involved. (They also added some coded messages to distinguish the copy from the original, challenging other scientists to break the code. They did, fast.) Then they removed the DNA from a related bacterium and replaced it with this synthetic genome. The resulting bacterium was then able to replicate, proving that the replacement genome worked.

The achievement gained worldwide publicity as the creation of the first synthetic life form – but that is an exaggeration. It's like overwriting part of your computer's memory with an exact copy of the same code, typed in by hand, and claiming to have built a new computer. The manufacture of Synthia was also condemned as 'playing God', an equally exaggerated criticism.

Synthia is important, though not to the extent that the hype suggested. It shows that long DNA sequences can be assembled from scratch in the laboratory. It adds weight to the belief that the activity of DNA in an organism follows from the laws of chemistry, rather than some mysterious aspect of life. And it is a useful step towards the Minimal Genome Project, whose objective is to make a synthetic bacterium with the smallest genome that allows it to replicate.[9] This hoped-for bacterium has been dubbed *Mycoplasma laboratorium*, and unlike Synthia, its genome will not be a copy of an existing natural one. But the rest of the cell's biochemical machinery will still be taken from a pre-existing organism.

That would be like writing a new operating system and loading it into an existing computer: closer to making a brand new computer from scratch, but not there yet. So genuine synthetic life is still some way off.

18 Is Anybody Out There?

· ·

Armed with an understanding of the different meanings that might
be assigned to the term 'life', we can now return to the question
raised at the beginning of the previous chapter: Does life exist
outside our own planet?

Scientists have never observed an alien life form, except perhaps
some tiny fossils in a meteorite, found in Antarctica and designated
ALH 84001, which some scientists think are evidence of past life on
Mars (Figure 74, see over). In 1996, NASA scientists announced that
the meteorite, thought to come from Mars, contained tiny fossil
bacteria. That claim remains controversial, and seemed to have
been comprehensively demolished until a recent reappraisal left a
tiny bit of room for hope, by answering some of the original
objections to a biological origin.[1] It's pretty clear that the meteorite
did come from Mars. Trapped gases in tiny bubbles in the rock
match the profile of Mars's atmosphere extremely well, and
calculations indicate that rock could have been blasted out of the
Martian surface by an impact with a small asteroid. And if that
happened, some of the debris from the blast could have ended up
impacting the Earth and landing on the ground in the Antarctic,
where this particular meteorite was found. The rock does contain
strange, tiny shapes, but whether these shapes were once alive is
the key issue – many think that the 'fossils' might be the result of
non-biological processes. It's difficult to give a definitive answer.
Since extraordinary claims require extraordinary evidence, the
burden of proof is on those who assert that the shapes are fossils of
once-living organisms.

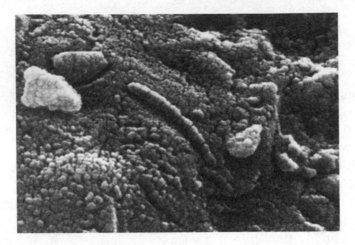

Fig 74 Suspected fossil bacterium in a meteorite from Mars. It is less than one-thousandth of a millimetre long.

Can we say anything genuinely scientific about life on other planets?

. . . .

I think we can, thanks yet again to the role of scientific inference.

Our current understanding of the origins of Earthly life suggests that there is nothing particularly special about our planet, so we should expect to find life elsewhere. Even intelligent life, aliens. Some scientists disagree, and argue that the Earth *is* special, so that life elsewhere may be very unusual, and complex life very unusual indeed. In their book *Rare Earth*, Peter Ward and Donald Brownlee make a persuasive case for this assertion, listing numerous features of the Earth and the Solar System that make it particularly suitable for life.[2] They accept that life elsewhere is entirely possible, but expect it mostly to be at the level of bacteria. Intelligence, they argue, will be very rare indeed. Others go even further, and assert that our planet is unique: the only place in the entire, vast universe where life exists.

Scientific opinions on the prospects for alien life take one of three broad positions:

- Alien life does not exist (not by definition, but by sensible scientific principles).
- Alien life does exist but must be very like terrestrial life.

- Alien life does exist and much of it is totally unlike terrestrial life.

The first position is rather negative, but since there is currently no convincing evidence of alien life, it is safe from attack, at least for now. However, current scientific understanding of the origins of life by natural physical and chemical processes does not limit life to one planet among 200 billion galaxies, each having on average up to 400 billion stars, a significant proportion of which (maybe one in four) probably has several planets. So it would be a big surprise if the Earth really is the only place in the universe where life exists, even if it has to be exactly like ours.

The second position is the most respectable scientifically, though not the most imaginative. We know that life forms like ours are definitely possible, and we don't know for sure that anything different can occur naturally. (Unnaturally is another matter.) The emerging science of astrobiology, or exobiology, combines Earthly biology with astronomy. Until recently it was almost solely focused on the prospects for Earth-like life, requiring an Earth-like planet. The more we understand about the origins of life on Earth, the more stringent these requirements become, and our best estimate of the chance of finding such kinds of life is correspondingly reduced.

However, the third position is slowly gaining ground. There are many valid scientific reasons to think that alien life need not be exactly like ours. One of the most important characteristics of life is that it is adapted to its environment – the basic feature of evolution. There seems to be no good reason why organisms could not evolve in, and become just as adapted to, an environment that differs from anything found on Earth. Insisting on an Earth-like environment as a prerequisite for life seems too narrow. It would be like Victorian explorers expecting all human beings to resemble Victorians in dress, manners and social structure, and ruling out the African forests as a human habitat on the ground that there aren't any milliners' shops there. Many astrobiologists are coming round to the view that while other Earths (just like ours) may indeed be rare, alien life might exist on worlds that differ from ours. A better term for this view is xenobiology, the biology of strange life; 'xenoscience' might be even better,[3] to make the point that alien 'biology' might be radically different from anything in our biology textbooks.

. . . .

How likely is it that the Earth is the *only* planet, anywhere in the universe, with intelligent life?

To keep the numbers simple, assume there are 10^{22} planets. By the law of large numbers, in order for there to be one planet on average with intelligent life, the probability of a planet harbouring intelligence must be 10^{-22}, one chance in ten sextillion. If the probability is 100 times bigger, then we expect 100 such planets; if it's 100 times smaller, we expect 1/100 such planets – which I'm inclined to interpret as 'none'. It therefore requires some very precise cosmological fine-tuning to hit the magic number 1 that makes our own world unique. It seems unlikely that any plausible physical mechanism could translate the probability of intelligent life into a specific number of planets. So either Earth won the jackpot in a cosmic lottery, or we are not alone.

Standard calculations indicate that in the critical case when the probability is exactly 10^{-22}, the probability of intelligent life being unique is 37%. The probability of *no* planets with intelligent life is also 37%, and the probability of two or more is 26%.[4] Those aren't bad odds, but it is a sobering thought that even when the universe is exquisitely fine-tuned for humans to be unique, we should expect no worlds with intelligent beings just as often as a unique one. And more than one is almost as likely.

If some day we discover that we really are alone, then either we'll need a better mathematical model, or we'll be forced to conclude that some kind of cosmic destiny has arranged for us to be unique. Right now, the best guess is that we are not alone. Planets with intelligent life are probably rare, but the universe is so vast that if there were about a quadrillion such worlds in the universe, all currently harbouring intelligent life,[5] then there would be less than ten thousand in our own galaxy. On average, the nearest one would be about a thousand light years away. So the universe could be teeming with life, and we'd never encounter it.

. . . .

A quick look on the Internet reveals that many non-scientists are convinced not only that aliens exist, but that they have already paid us a visit. However, the US Government has instituted a cover-up, so we don't see aliens walking around our neighbourhood. I'm willing to believe that the US Government – or any government – *could* decide to cover up alien visitations; however, I doubt it could succeed for long, and I don't think it has – because I don't think there's anything to cover up.

The main reason for disbelieving these tales of alien visitations is not political opinion, or the mindset of the believers, but the science of alien life – which is thriving, and entirely scientific, despite a complete absence of observations of aliens.

It is possible to do good science, even when the subject under discussion has never been observed. Physicists have done a huge amount of work on the Higgs boson, a particle predicted by the 'standard model' of subatomic particles, but no one has yet observed one. In fact, that's what the Large Hadron Collider, which famously broke down days after it was first switched on, is looking for. Assorted nations have committed $9 billion to get it built and keep it running. It is now back in action, but unlikely to find the Higgs boson before this book is published.

No one has ever observed a superstring, but string theorists have devoted a lot of effort to these hypothetical objects because they have the potential to unify quantum theory and relativity. No one has observed a universe coming into existence, but on the whole, cosmologists don't resign and get jobs as hedge fund managers because of that. No one has observed the interior of a black hole, the birth of a Neanderthal, the emergence of life on land, a herd of wandering sauropods or the gravitational field of the Andromeda Galaxy. Not directly, and in many cases not at all. But these areas are all solid parts of the scientific enterprise.

In fact, you could do a lot of excellent science in a quest to prove that aliens *don't* exist. *Rare Earth* is a case in point.

Science is not simply a matter of direct observation. It is an intricate interplay between theory and experiment, and the experiments are often indirect. The strength of science is inference. No one alive in the past million years observed the evolutionary divergence of humans from chimpanzees, but scientists are in no doubt that this event happened, because many independent lines of investigation all point inevitably to that conclusion. The fossil evidence, the observed ages of the rocks that contain them, the immensely detailed DNA evidence, and the biochemistry of chimp and human bodies all make this contention at least as certain as the Earth going round the Sun.

Fig 75 Picture alleged to be of a 'grey' alien.

Martian microfossils aside, the main claims of alien observations come from ufologists, New Agers and believers in the paranormal. Abduction is a common scenario. The typical ufological alien, often called a 'grey', is humanoid, shorter than us, with a greyish skin, a large head, huge oval eyes and tiny nostrils (see Figure 75). Its bodily proportions are different from the human, and its limbs have different joints. Greys dominate reports of encounters with aliens to an extent that is quite remarkable: 90% in Canada, 65% in Brazil and 40% in the USA. In Europe the figure drops to around 20%, and Britons, eccentric to the last, reach only 12%.[6]

Ironically, the main reason for disbelieving such reports of alien visitors is not that greys are too strange to be credible. It is the exact opposite: they are not strange enough. Greys are *the wrong kind of alien*. They are far too similar to us. And the science that justifies that claim is good, solid, terrestrial biology.

From the 1960s, my biologist friend Jack Cohen gave over 300 lectures to schools about life on other planets. One of the key scientific principles he explained to them was how to decide which features of life on our planet are likely to be found in alien life forms, if they exist, and which features are merely accidents of evolution on this world and would not be expected elsewhere. I say 'likely to' because the discussion has to be theoretical at this stage in our exploration of the universe.

Cohen distinguished these two types of feature, calling them universals and parochials. The words can be used both as adjectives ('this is a parochial feature') and as nouns ('this feature is a parochial'). Five digits on a hand is a parochial, but appendages that can manipulate objects are universal. Wings covered in feathers are parochial, but the ability to fly in an atmosphere is a universal. Daisies are parochial, but photosynthesis – obtaining energy from light – is universal. The term 'universal' does not mean that such creatures will exist everywhere, not even on every suitable planet. Flight needs an atmosphere, for instance, but we don't expect every planet with an atmosphere to have flying creatures. Universals are features that are very likely to evolve on other suitable worlds. Parochials, on the other hand, are local accidents, and we would not expect to see them elsewhere.

The creatures of our planet, at any level of detail, are mostly parochial instances of universals. Each particular type of eye – and there are hundreds of clearly different structures – is a parochial, but vision is a universal. Legs differ from one creature to the next, but locomotion is a universal. Such examples motivate a test that can distinguish the two types of feature. Did the feature evolve just once, or many times independently? 'Just once' allows for subsequent modification in many different descendant species. 'Independently' means that the different instances have no such evolutionary connection.

In humans, the airway and foodway cross, resulting in many deaths each year from choking. We share this structure with most mammals. It's a very poor 'design', and goes back to an evolutionary accident. About 350 million years ago there were no large land creatures, but in the seas and oceans were many fish. Some had lungs on the top of their bodies, some had lungs underneath. (Taking in some of the planet's atmosphere to power chemical reactions is universal: where the associated organs go is parochial.) The lobe-finned fishes that evolved into land animals, mentioned as transitional forms in Chapter 5, happened to have their lungs underneath. Our own awkward arrangement is a consequence. It is reasonable to think that some other fish might have made the evolutionary transition to land instead, even on this planet; certainly there seems to be little to stop an analogous event happening elsewhere. So foodway-crossing-airway is a parochial.

Fossil evidence supports the view that the transition from sea to land was gradual. Fish did not suddenly come out on land, contrary to the Gary Larsen 'Far Side' cartoon in which three young baseball-playing fish are gazing wistfully at the ball, which has popped out onto the land. The transition from fins to limbs probably happened while the fish were still in the sea, scuttling around in the shallows and increasingly using their fins to push against the mud and sand. Their ability to 'walk' along the seabed evolved in concert with the structure of their limbs. By the time they were inhabiting the land permanently, their fins had changed to legs. Pentadactyl limbs, ending in five digits, evolved during this period, and were transmitted from the fishes that scuttled in the shallows to amphibians, then to reptiles, then to mammals and us. The entire structure of our limb joints is parochial. However, possessing jointed limbs is universal. It evolved independently in insects, for example.

The 'evolved here many times' test for a universal is closely linked to Earth's particular evolutionary history. Intelligence is a universal by that definition: it has evolved in the octopus and the mantis shrimp, as well as mammals. But human-level intelligence – what Cohen and I call 'extelligence', the ability to store cultural capital and know-how outside ourselves in a form that can be widely accessed – seems to have evolved only once on Earth. Dolphins are smart but they don't have libraries. So extelligence fails the evolutionary test for being a universal. However, it's reasonable to argue that it *ought* to be a universal. Our specific brain structure is parochial, and even brains as such may be, but extelligence is a generic trick, offering clear evolutionary advantages, rather than a specific accident of heredity. Perhaps we haven't waited long enough for it to arise again.

This suggests that we should broaden our definition, replacing it by a more theoretical version. A parochial is a specific feature that appears to have arisen by accident, and would be unlikely to occur in the same form in a rerun of Earthly evolution. A universal is a general feature that could be realised in many different ways and appears to offer a clear evolutionary advantage, and would probably emerge again if evolution were to be reset and run a second time. Agreed, this definition leaves room for debate, but the whole idea is intended as a guideline, not as a hard and fast rule. It is a guideline for *how to think* about the possibility of life on other planets.

How good a guideline it is will depend on whether we find such life, and what it looks like if we do.

. . . .

Most of us absorb our images of aliens from television and movies. Mine came initially from the *Eagle* comic, and the exploits of Dan Dare, Pilot of the Future, tangling with green-faced Venusian Treens and the evil Mekon, who had an enormous head and a small body, and rode on some sort of antigravity cushion. Comics had more room for imagination than movies or TV. In those media, before computer graphics reached its current levels of realism, aliens had to be thinly disguised human actors, insects magnified to gigantic proportions, or invisible presences that glowed in the dark, emitted sparks or disturbed the air and moved the curtains. Now they can be impressively detailed creatures that inspire terror, like the mother alien in *Alien*, or they can be cute and cuddly like the Ewoks in *Return of the Jedi*. And that is what their designers intend, and it is why the aliens in the media mislead us about what real aliens might look like.

Media aliens are invented in order to stimulate specific human emotions. This makes them hopelessly parochial, and many of their features make no scientific sense at all. The aliens in *Alien* grew inside a human body until they reached the size of a cat, and then burst out gorily through the chest wall. Leaving aside the question of why the victims often didn't know they had a cat-sized lump inside them, how did a creature on another planet evolve to exploit the biochemistry of a human body? They might just possibly be generalist parasites that can make use of a variety of other creatures as hosts, but it is almost impossible for that kind of generalism to evolve. Parasites co-evolve with their hosts, and are usually very specialised. Dog fleas can't survive for long without a dog, even though they may temporarily infest a human.

The great universal is evolution. That is how life can diversify, and some of it becomes more complex, while growing ever more suited to the conditions on its home planet. A real alien would have evolved in some environment elsewhere in the universe, and it would be adapted to that environment. So, in order to invent scientifically credible movie aliens, you would have to invent a plausible environment and evolutionary history as well. Mother alien and her parasitic children don't work. But few movie or TV producers go to those lengths. 'It's science fiction, it doesn't have to make sense,' they seem to think. But this is a recipe for unconvincing entertainment, bad science fiction, and even worse science.

Another source of images for aliens is human culture. Previous generations saw – well, some thought they saw, and many believed in – ghosts and pixies and other supernatural creatures. Claims of alien visitations and abductions are part of a long tradition of fearsome supernatural creatures, presented in terms that made sense to the culture concerned. They all probably have the same origins: sleep paralysis, in which we can become awake while still in the dreaming state, where our limbs refuse to move and our critical faculties are suppressed. The dreamed abduction seems real because the part of the brain that distinguishes dreams from reality is not functioning, and we experience a feeling of terror because *we can't move.*

Specific images and aspects of these supernatural visitors spread through the culture. Even people who don't believe in UFOs 'know' that aliens have huge, dark eyes and big heads. That's how you tell they're aliens. Actually, it's how you tell they are fictions. Greys are too humanoid: they are built from minor variations on human parochials. They are the lazy way to invent an alien: make it like us but change a few features for dramatic effect. It's not just the shape – bipedal, head at the top, human-shaped skull (just exaggerated). Greys breathe our air – but few creatures that had evolved on another world would be able to do that, unless that world's atmosphere were very similar to ours. Even humans have trouble on this planet if they merely move to high altitude, where the air has a similar composition but is thinner. Only people who grew up at such heights are comfortable there. Go to Peru, and you'll see what I mean.

. . . .

That said, there is room for disagreement. There is also a long tradition of scientific theorising about aliens, arguing that they must be very much like us – maybe not in form, but in their biochemistry and the kind of environment they could live in. The fledgling science of astrobiology mostly takes what we know of Earthly biology and projects it onto the background of alien worlds which we know about through astronomy. The search for alien life then becomes the search for habitable worlds, where 'habitable' means that we could live there. Or something very similar to us, adapted to local conditions as Peruvians are to altitude.

If you think aliens should resemble us, you can assess the prospects for aliens by starting with life on Earth. How does Earthly life (identified simply with 'life') work? Despite its diversity, everything relies on DNA, RNA and a very standard system of molecular machinery. There are some variations, but they are slight. What do Earth's creatures need to survive? Water, oxygen, land to live on, comfortable temperatures, low levels of radiation. Energy from the Sun. A stable environment – well, fairly stable: not *too* many earthquakes, volcanic eruptions, tsunamis, forest fires, or inbound comets and asteroids.

Does this mean that all alien life forms would need the same? That they, too, would have the same kind of DNA? The argument for this view boils down to one simple fact: the only life that we know anything about exists on this planet. All else is hypothetical. From this point of view, the only sensible scientific position is that our kind of life is the only kind of life. Don't agree? Then show me.

If you insist. There are good reasons to suppose that DNA may not be the only game in town. In the previous chapter we saw that virtually every major player in the Earth's biochemistry can be modified, and still work. You can change the molecular structure of DNA. You can use a different genetic code to turn sequences of DNA bases into amino acids, the basic building blocks of proteins. You can even change the *number* of bases that encodes an amino acid, from the usual three to four. You can change the list of amino acids. You can use different proteins for specific functions. Life can exist without oxygen, without sunlight, and – according to a conference held at the Royal Society a few years ago – without water.

Forms of life that don't use carbon–oxygen chemistry as a

source of energy have to do something else, and one possibility uses sulphur and iron instead. Günter Wächtershäuser, a chemist and patent lawyer, suggested that life on Earth first arose in a hydrothermal vent on the ocean floor, exploiting the chemicals that are common in such places, notably iron sulphide.[7] With the aid of small quantities of catalysts such as nickel and cobalt, hot water flowing over iron sulphide can form reasonably complex organic molecules, known as metallo-peptides. Some experimental support exists for these ideas, but it is not clear just how complex this type of chemistry can become. It is, however, a plausible alternative to carbon chemistry, and might have given rise to primitive forms of life.

Less hypothetically, there is a lake at the bottom of the Mediterranean Sea, 3.5 kilometres down, west of Crete. I say 'lake' because a dense layer of unusually salty water has collected there, pooling on the seabed. It contains hardly any dissolved oxygen, but a lot of hydrogen sulphide, which oozes out from a thick layer of mud. The only life that ought to exist in the lake is anaerobic bacteria, which don't require oxygen. But in fact there are small, complex animals, with a hydrogen–sulphur metabolism. Bill Martin, an evolutionary biologist, believes that these animals change our view of the origin of eukaryotes.[8]

The orthodox theory is that eukaryotes evolved because of a massive build-up in the ocean and the atmosphere of oxygen, the waste product of photosynthetic bacteria and algae. Oxygen is a potential source of energy, so organisms could evolve to exploit it. Mitochondria, vital to eukaryotes, do just that; they also protect the cell against the toxic effects of oxygen (things *burn* in it). Martin argues that oxygen is reactive only when it is in the form of free radicals, but mitochondria *create* free radicals, so they make the problem worse. And extracting energy from oxygen is so complex that it must have taken a long time to evolve. So for billions of years the oceans would have been jam-packed with hydrogen sulphide, and the anaerobes would not have been poisoned by oxygen. So maybe it wasn't oxygen that led to the eukaryotes, but hydrogen and sulphur. The animals in the undersea lake may be relics of that process, though they will surely have been changed by more than a billion years of evolution. If Martin is right, Earthly life began as unearthly life. Since oxygen wasn't needed here, it's silly to think that it must be necessary for aliens.

There is another argument against the uniqueness of DNA and the associated chemistry. Agreed, there are millions of different species on planet Earth, all using the same biochemistry, but that doesn't imply that nothing else is possible, because all evolved from the same primitive ancestral forms. Life reproduces: as soon as anything works, you find it everywhere. In a sense, those millions of species provide no more compelling evidence for the necessity of a particular biochemical scheme that any single one of them would.

Not only that: even on this planet, life exists in wildly different habitats, so different that until recently biologists denied that life could ever exist in some of them. These life forms are primitive, on the level of bacteria. They are collectively known as extremophiles – organisms that can survive in extremes. Some live happily in boiling water, others in water that has been supercooled, below its normal freezing point. Some have been found three kilometres underground, some in the stratosphere, and some can survive radiation levels that would be fatal to all other life forms. Late in 2010, a team led by Stephen Giovannoni of Oregon State University drilled nearly 1,400 metres into the bed of the Atlantic Ocean, where they found bacteria thriving at a temperature of 102°C.[9] NASA scientists have reported that some bacteria in a Californian lake use arsenic – poisonous to most organisms – in place of the usual phosphorus, though this finding is controversial.

The term 'extremophile' reflects unconscious human bias. To a creature that lives in boiling water, it is *we* who are occupying an extreme environment. Somehow we survive the appalling cold of a British summer. The word 'survive' suggests difficulty, but an extremophile is not clinging desperately to the edge of survival in its boiling hot pool: it is comfortable there, and would die if it moved into water that was merely scaldingly hot. The same goes for the other class of extremophile, able to live in freezing cold. Our environment would normally be far too warm for it.

I find it strange that both boiling hot and freezing cold are somehow lumped together into the single category 'extreme', but the stuff in between is different. It smacks of parochialism, and reminds me all too closely of Goldilocks and the Three Bears.

I'll come back to that.

. . . .

Not so long ago, many people, including scientists, had what they thought was a very good reason to believe that the Earth is unique, the only planet that can support life. Their reasoning went much further: they 'knew' that the Earth was the only planet in the Solar System that could support life, because the Solar System was the only place in the universe that had planets. In support of this view was a clear statement of fact: planets round other stars had never been observed, so their very existence was hypothetical. Hypotheticals have no place in solid science: the only reason to think that other planets might exist was pure speculation, based on our limited knowledge about the formation of the Solar System.

To be sure, this suggested the exact opposite: that there is nothing very special about the Sun, so similar processes have probably occurred elsewhere. That meant planets. It was plausible, but there was no proof, so it wasn't science.

This particular line of thinking has gone the way of the dodo. As I write, we know of 518 planets circling other stars[10]: the technical term is 'exoplanets'. More are being discovered every week. It is becoming obvious that a significant proportion of the stars in the universe have planets; possibly most of them. We may never be able to observe the bulk of these worlds directly, but a random sample usually represents a wider truth. Planets are no longer the issue. The only reason we didn't observe them earlier was that we lacked the technology to detect them. So the frontier of the debate has retreated to the existence of *Earth-like* planets. Almost all known exoplanets are huge, bigger than anything in the Solar System, dwarfing even Jupiter. The diehards retreated to a previously prepared position, now insisting that the evidence merely proves the existence of gigantic planets that could scarcely be more different from Earth – which to them means no possibility of life on such worlds. Again, there is actually a good reason why most of the known exoplanets do not resemble Earth: the methods used to detect exoplanets work best when the planet is very large.

Improved observation techniques have pushed that frontier back too: we now know of much smaller exoplanets, and can already detect the main gases in their atmospheres. In 2008 Mark Swain's team at the Jet Propulsion Laboratory in California made the first detection of an organic molecule, methane, on an exoplanet.[11] They found it on HD 189733b, a 'hot Jupiter' about 63 light years from Earth. Water vapour has been found on GJ 1214b,[12] and the same methods should be able to find oxygen.

This suggests that it is unwise to dismiss reasonable possibilities on the ground that there are no observations to support them. You need independent evidence that those possibilities are intrinsically unlikely. A lack of observations can change at any moment with the invention of a new technique. So the current absence of evidence of alien life forms that differ from those on Earth may simply be due to the absence of evidence of alien life forms. Just as the absence of evidence of Earth-like planets was, until recently, due to the absence of evidence of planets – which was not evidence for the absence of planets, Earth-like or not.

• • • • • • • • • • • • •

The case that complex life elsewhere in the Galaxy, or the universe, is very uncommon has been made, eloquently and comprehensively, in *Rare Earth*. Ward and Brownlee list a large number of special features of our planet, all supposedly necessary for life to exist, and then work out how likely such a combination of features is. Their result is: very unlikely indeed. They don't rule out simple life forms, such as bacteria, but they argue persuasively that anything even as complex as, say, a goldfish, must be a very rare thing in our universe. They don't claim that Earth is unique in having such creatures, but other Earths, if they exist at all, will be very thinly spread.

I'll list some examples of these features – they are based on solid science, much of it surprising and recent, and of interest in its own right. I'll restrict attention to three astronomical ones. The first two are relatively new; the third is much older.

1. Jupiter protects the inner planets, Earth among them, from being bombarded by comets. A dramatic instance of this process was the break-up of comet Shoemaker–Levy 9 in 1994. Earlier, the comet had swung close to Jupiter, and was diverted in its orbit so that it would return after a few years. As it approached the giant planet, it broke into twenty pieces, which slammed into Jupiter releasing the energy of six million megatons of TNT – roughly six hundred times the world's total store of nuclear weapons. If any one of those fragments had hit the Earth,

nothing higher than bacteria would have survived, and probably not even them. Without Jupiter, comets would be hitting the Earth every twenty years or so.

2. The Earth's Moon keeps our planet's axis of rotation stable. Mathematical calculations show that a world lacking a moon that is large compared with itself will suffer erratic changes in the direction of its axis over periods of tens of millions of years.[13] Such large moons are uncommon; it is thought that ours originated from a massive collision between Earth and a body the size of Mars during the early stages of the formation of the Solar System. Such collisions are rare.

3. The Earth is situated within the Sun's habitable zone: a hollow shell of space inside which liquid water can exist on a planet's surface. Get too close to the Sun, and water will turn to steam and may boil away entirely; too distant, and it will freeze. The habitable zone is limited: Mercury and Venus, close to the Sun, are on the inside of it; Mars, Jupiter, Saturn, Uranus and Neptune are outside it. We got lucky.

Rare Earth lists several dozen such features, and they are regularly trotted out in television science programmes as proof that Earth is very close to unique. However, the case for Earth's rarity, like premature news of Mark Twain's death, has been greatly exaggerated.

Indeed, the importance of each item in the list has been exaggerated. Worse, the significance of any such list has been exaggerated. I'll go through the three items above in turn, and then turn to my more general objection.

Jupiter. As Shoemaker–Levy 9 shows, there are occasions when Jupiter does indeed protect the Earth from comets. But that does not imply that it always has a beneficial effect. It can also divert incoming comets, causing one that might have missed the Earth to hit.

World governments and NASA are starting to worry about NEOs: near-Earth objects. These are lumps of cosmic rock whose orbit round the Sun can bring them close to Earth. Many are asteroids, bodies ranging in size from a tennis ball to one-third of the diameter of the Moon, although the largest bodies in near-Earth orbits today are much smaller than that. Asteroids are found by their thousands in the asteroid belt between Mars and Jupiter.

Indeed, most Earth-crossing asteroids probably originated in the asteroid belt. If they had stayed there, they would pose no danger to our world. What made their orbits change, to become Earth-crossing?

Jupiter.

Mathematical calculations show that Jupiter has a big effect on asteroid orbits. In fact, as the most massive planet in the Solar System, Jupiter has a big effect on the orbits of small bodies. One of the things that Jupiter can do is disturb suitable asteroids, causing their orbits to elongate until they cross the orbit of Mars. This makes them Mars-crossing, not Earth-crossing. But now, if they come close to Mars, they can be diverted again, and now their orbits can cross the orbit of our own Blue World. So Jupiter centres the ball, and Mars scores.

Jupiter has two faces. One is that of protector. The other hurls rocks at us.

Ward and Brownlee discuss this, and argue that the occasional asteroid impact may be good for evolution, shaking up the biosphere. And so it might, but I wonder why asteroids are beneficial in this respect, while comets are not. It seems like special pleading. In fact, the presence of Jupiter may do more harm than good.

The Moon. Unlike most astronomers, I'm not convinced that the current theory of the origin of the Moon is correct,[14] but whatever the mechanism was, it does seem likely that satellites whose size is comparable to that of their primary planet are quite rare. So I'll concede that. And I agree that the presence of such a body does stabilise the axial tilt. However, it is not at all clear that if a planet's axis changes its direction over a period of tens of millions of years, which is what the mathematics says, then this poses an insuperable problem for evolution. Earth's creatures have coped with ice ages that come and go every ten or twenty thousand years, which is far more rapid than a change in axial tilt. Land creatures can move as the climate shifts – we're talking a few hundred metres per year, and they're moving faster than that today in response to climate change – unless they run out of land, which can happen. (Our elderly cat moves faster than that when chasing a mouse, but I'm referring to changes in average geographical location.) Birds can fly across open sea. And ocean creatures *wouldn't notice any difference.* Since it is generally agreed that life began in the Earth's oceans, and got very complex there, then the tilt of the Earth's axis doesn't matter a hoot.

Habitable zone. The habitable zone of a star is often referred to as the Goldilocks zone because it's 'just right'. The problem with habitable zone ideas is not that the concept is completely silly, but that it is too simplistic. It is, for example, not at all clear whether the Earth is within the Sun's habitable zone. An airless Earth might have boiling hot surface temperatures, like the Moon when the Sun is overhead; on the other hand, one with too little carbon dioxide or a white reflecting surface might be covered in ice, as some of the planet is *now*, and most of it was during the period of Snowball Earth about 700 million years ago. Both Mars and Venus might support liquid water in suitable circumstances, but in those that currently prevail, Mars can get down to 15 degrees below zero and Venus is hot enough to melt lead. The highest surface temperature recorded on Mars to date is 27°C, but only on rare summer days.

It's worth taking a closer look at the mathematics of habitable zones, to see where the difficulties arise. The calculations start from a central idea in the physics of heat, called a black body. A green object looks green, to the human visual system, because the object reflects sunlight in a range of wavelengths that our brains interpret as 'green'. A black object does not reflect any wavelengths in the visible range; black is the brain's default for such objects. A physicist's black body is an idealised and extreme version of this: it reflects no electromagnetic radiation whatsoever.

However, reflection is not the only way for an object to emit radiation. A black body at a temperature of zero degrees kelvin – 'absolute zero', the lowest possible temperature – would emit no radiation of any kind. But at any other temperature, a black body does emit radiation; it just doesn't do this by reflection. Instead, it glows incandescently, like a red-hot iron bar. The intensity of radiation emitted depends on the temperature and the wavelength of the radiation. Classical physics predicts that a black body should emit an infinite amount of energy, but that makes no sense. In 1901 Max Planck derived a new formula that agreed with observations, and this was later interpreted as evidence for a quantum world.

Planck's law can be used to derive a formula for the temperature of a planet orbiting a star, and that lets us calculate where the inner and outer edges of the habitable zone are. There are two versions of the formula. The simplest one models the planet as a black body. However, a real planet will reflect some of the radiation that hits it, and the second model takes this into account by incorporating an extra quantity into the formula. It is called the albedo of the planet, which is the fraction of incoming radiation that is reflected away.

Only the first version can yield a habitable zone that depends only on features of the star. As soon as albedo comes into play, the habitable zone also depends on features of the planet – real or hypothetical – that is under consideration. The second version is more general than the first: if we set the albedo to zero, the value for a black body, we recover the first model. The formula relates the planet's temperature to the star's size, its surface temperature, the distance from the star to the planet, and the planet's albedo.[15]

First, let's calculate what the Earth's temperature would be if its albedo were zero, the value for a black body. The answer is 279 K, or 6°C, placing us just inside the habitable zone. However, if we use the observed albedo, which is 0.3, the temperature becomes 254 K, which is –19°C – well below the freezing point of water. So assuming the correct albedo leads to the paradoxical result that the only known habitable planet in the universe does not lie inside its star's habitable zone.

To locate the outer edge of the habitable zone, we consider a hypothetical planet whose surface temperature is the freezing point of water, 273 K. Then we solve the equation to derive the distance. For the inner edge, we do the same thing, but using the temperature of boiling water, 373 K. Again, there are two versions. For an albedo of zero, the value for a black body, the Sun's habitable zone extends from 83 million to 156 million kilometres. If we set the albedo to 0.3, the measured value for the Earth, then the habitable zone stretches from 69 million to 130 million kilometres.

The average distances of the four inner planets from the Sun, in kilometres, are 58 million for Mercury, 108 million for Venus, 150 million for Earth, and 228 million for Mars. So for albedo 0 the Earth just scrapes inside the Sun's habitable zone ... but so does Venus. For albedo 0.3, only Venus lies inside the habitable zone. The Earth and Mars are too cold, Mercury too hot.

Why, then, is the Earth habitable? Because its atmosphere contains greenhouse gases, mainly carbon dioxide and water vapour, which trap incoming radiation and make it warmer than it would be if no atmosphere were present. But the usual concept of a habitable zone does not take the planet's atmosphere into account. The idea that a *star* has a habitable zone, independent of properties of the relevant planet, is an oversimplification. Of course, in a broad qualitative sense it is true that if a planet is too near its star then any water present on its surface will boil, and if it's too far away the water will freeze. But 'habitable zone' lends a misleading air of precision.

Greenhouse warming is just one of a huge variety of effects that between them pretty much demolish 'habitable zone' as a useful concept. The surface temperature of a planet depends on many factors, only one of which is how far it is from its star, for a given heat output. For example, clouds and ice can increase the albedo, cooling the planet; so can sulphur dioxide. Carbon dioxide, methane and water vapour can warm it. Feedback loops between different factors further complicate the possibilities: warming seas can create clouds that reflect heat and light back, decreasing ice cover can allow more heat and light in.

Even taking all this into account, it is not true that the only place where liquid water can exist is on the surface of a planet in the habitable zone. For example, it used to be thought that Mercury was locked in a spin–orbit resonance, rotating once during the same time it took to revolve once round the Sun. If so, the same side would always face the Sun, just as the same side of the Moon always faces the Earth (give or take a bit of wobbling, known as libration). In fact Mercury does not do this, but there's every reason to expect that some worlds somewhere in the Galaxy might be very close to their star – much closer than the habitable zone defined above – and locked in such a resonance. In fact, this is the case for at least one exoplanet.[16] If so, one side of the planet would be very hot, the other side very cold ... and in between there would be a belt with more moderate temperatures, suitable for liquid water to exist. 'Just right', in fact.

Astronomers are almost certain that liquid water exists on a number of bodies in our Solar System that are well outside the Sun's habitable zone. Paramount among these is Europa, a satellite of Jupiter. There is convincing evidence that Europa, one-quarter the diameter of the Earth, has an ocean that contains as much water as all of Earth's oceans put together. Yet Europa's surface is solid ice. So where is the ocean?

Under the ice.

Measurements of Europa's magnetic field reveal changes that currently seem to be consistent with only one thing: a world-girdling ocean upon whose surface the ice floats. The water is kept warm by heat generated in Europa's core, probably caused by repeated squeezing by Jupiter's huge and powerful gravitational field. Jupiter's inner three major satellites, Io, Europa and Ganymede, are trapped in an orbital resonance: while Io goes round four times, Europa goes round twice and Ganymede once. This creates unavoidable tidal forces, and squeezing causes friction, which heats the core. This is not such an outlandish scenario: the Earth's continents and seabed of solid rock float on a vast underground ocean of magma.[17]

Europa is not alone in having such an ocean. Ganymede and Callisto may have one too, Io probably has one but it's sulphur, not water, and Saturn's moon Titan may have a subsurface ocean of slushy liquid methane.

Finally, there is the obvious point that the Earth's own extremophiles live in conditions that are outside the habitable zone: water at temperatures above its normal boiling point, and below its normal freezing point. Not far outside, but outside all the same. Could they have *evolved* in such extreme conditions? That's less clear, but we've already seen that a plausible theory of the origin of life has it evolving first as . . . extremophiles.

. . . .

Protective gas giants, stabilising moons, and Goldilocks orbits . . . *Rare Earth* lists dozens of such factors, and like those three, most of them are open to serious challenge. But there is a more general issue, a mathematical point: logic.

It is all very well to list dozens of special features of the Earth, all of which definitely played a significant role in the evolution of life. But it is wrong to conclude from this (alone) that those features are *necessary* for life. The correct conclusion is that they were *sufficient*. 'Sufficient' means that with them, life arose. 'Necessary' means that without them, it would not have arisen. The two are different, and it is the first that the list of features supports. For you to get wet, it is sufficient to stand outside in the rain without protection. But that's not necessary. You can fall in a lake or take a bath instead.

Evolution is a universal, and its main feature is that creatures evolve to suit their habitat. If some form of life can exist in some habitat, even if it seems hostile to us, then life can evolve to do so. It doesn't care about our opinions, because we're not going to be living there. If we approach questions about alien life with the tacit assumption that the only sensible form of life is us, we will ignore all the other possibilities. The word 'extremophile' is human-centred: it starts from where we are, and declares that to be what's sensible and reasonable. The further away we go from our self-defined centre, the more 'extreme' things become.

I remember a museum exhibit about deep-sea fish, which said something to the effect that 'their strange shapes reflect the strange conditions under which they live'. It seems to make sense: strange conditions imply strange shapes. Not like normal conditions, which imply normal shapes. Like us. But it's all back to front. Normal conditions, in this sense, are the ones we are accustomed to. So are normal shapes. But we are as different from the fish as they are from us, in both shape and habitat. To them, we would be strange and they would be normal.

To evolution, we would both be normal – relative to our habitat.

. . . .

A more imaginative reading of the Goldilocks tale makes the same point, and raises a far more interesting set of questions. Mummy Bear's wimpy porridge was too cold for Goldilocks, and Daddy Bear's macho male porridge was too hot, while Baby Bear's intermediate porridge was just right. And so they were – for Goldilocks.

For Mummy Bear, however, the intermediate porridge was too warm. For Daddy Bear, it was too cold. Goldilocks' point of view is not privileged. Woolly-minded social relativism though it may be, I think that Mummy and Daddy Bear both had valid opinions too.

Discussions of such things as Jupiter's alleged importance in protecting the Earth from comet impacts often run along the following lines: 'Without Jupiter, the Earth would be hit by a comet every twenty years.' There's a sense in which such statements are true, but a closer look reveals that they don't address anything of substance. They're like sports commentators saying, 'If only he hadn't been offside, that goal he just scored would have won the match.' But if the player had not been offside, he would have been in a different place. To score a goal, he would have to have kicked the ball differently. You can't just change one factor, offside, and keep the rest exactly as before.

It's the same with Jupiter. Yes, if you took the Solar System as it is today, and magically spirited Jupiter away, comets would rain in upon an unprotected Earth. But if the Solar System had evolved without Jupiter, it would not be the same – in all other respects – as it is. It would have been quite different. More comets would have hit the Earth in the past, for instance, leaving fewer to hit it now.

The mathematics of many bodies moving under gravity, called celestial mechanics, is revealing an unsuspected aspect of planetary systems. Namely, that they *are* systems. Over billions of years, they organise themselves in complex ways. The biggest planets, the Jupiter-like gas giants, have the biggest influence. Other, lesser worlds, and even those only slightly less massive, get rearranged until the entire system fits together and acts as a whole. This is *celestial* Gaia.

Very recently, it has been discovered that Jupiter's influence on the Solar System has created a kind of celestial subway, a network of gravitational 'tubes' that can be perceived mathematically but consist of empty space.[18] These tubes are pathways along which matter can move more efficiently. The arrangement has come about as a result of subtle feedback effects, caused by gravitation. The equations for gravity are nonlinear, meaning that effects are not proportional to causes. Nonlinear systems have a tendency to behave in surprisingly complicated ways, and they tend to organise themselves by settling into special forms of behaviour.

Considering a Solar System without Jupiter, and arguing that it wouldn't be so hospitable for life, makes the same mistake as the sports commentator. It forgets that if you change one thing, you change *everything*. The evidence to date shows that most solar systems have huge planets like Jupiter. It seems likely that most of them also have smaller planets, though these are very difficult to spot at the moment. If so, then the Jupiters will organise their lesser brethren, and often enough there will be a few small worlds closer to the star, and some big ones further out. So even if it is indeed true that planets like Jupiter provide overall protection against comets and the like, it is no huge coincidence if one of them exists in roughly the right orbit. Nature does not simply build solar systems by plonking down planets at random. They have a self-consistent structure.

This is not to say that habitable planets are inevitable. There are many ways to fail to be habitable. But there are many stars – at least 4×10^{22} of them – in the universe, and probably even more planets. There are many ways to be habitable too, and habitable planets will not all be carbon copies of the Earth. Rerun the Solar System again, from different beginnings, and there's a fair chance that at least one world might still be suitable for life.

At some point in its history. Mars may have been suitable for Earth-like life, a billion or more years ago. In fact, it has been suggested that Earthly life was originally seeded from Mars. The current consensus is that this is probably wrong, but it's not totally out of the question. It will take a close look at Mars to decide the matter.

. . . .

In contrast to the *Rare Earth* story, I will describe some mathematical simulations and models devised by Harvard astrophysicists Dimitar Sasselov, Diana Valencia and Richard J. O'Connell, which suggest that planets capable of supporting Earth-like life may be far more common than has previously been thought.[19] Their results also call into question the common view that Earth is the ideal kind of world for the kind of life that we find here.

The starting point for their work is the realisation that the conditions that make our kind of life possible do not necessarily require the planets concerned to be of a similar size to our own. What matters is that they should resemble our own world in one key respect: the occurrence of plate tectonics. There is a growing suspicion that the dynamic movement of continents helps to stabilise the Earth's climate. In particular, carbon dioxide is recycled from the atmosphere to the ocean floor, where it is taken up by marine microorganisms and turned into carbonate; then the subducted carbonate is turned back into carbon dioxide by volcanoes. A stable climate helps liquid water to exist for geologically long periods of time, and water is required for our kind of life, even if other kinds might exist without it. That, in turn, enables evolution to generate complex water-dependent life forms.

It had been assumed that plate tectonics is rare, and that it requires a world of comparable size to our own. The crust of a much smaller world could not break up into suitable plates, while a much larger world would be a gas giant and not have a surface as such anyway. Sasselov and colleagues have shown that both these assumptions are false. Plate tectonics may actually be very common, and could occur on planets much larger than the Earth. The reason is the possibility of 'super-Earths': rocky worlds with a similar geological composition to ours, but having much greater mass. No one had previously investigated the internal geological processes of such a world, probably because no such exoplanets were then known. Indeed, virtually all known exoplanets were so large that they had to be gas giants, and this was still the case when the team began their modelling and published their first paper.

By 2005, however, the picture had already started to change with the discovery of the exoplanet GJ 876d, which orbits the star Gliese 876. This was smaller than the typical gas giant exoplanets then known, though still much larger than the Earth; there were hints that it might be mostly rock, rather than gas. However, there was no good way to measure the planet's density, which would decide the issue, because the only known method required the planet to cross the face of its parent star when viewed from Earth. In 2009 a new exoplanet was found, CoRoT-7b, which did cross the face of its star. Now a density estimate was feasible, and the result was definitive: CoRoT-7b is made from rock. It has about 4.8 times the mass of the Earth and 1.7 times its radius. By 2010 a second super-Earth that transits its star had been located, known as GJ 1214b, with a density closer to that of water than rock, suggesting that it has a thick gaseous atmosphere. This planet has 6.5 times the mass of the Earth and 2.7 times its radius.

Now there were real planets to supplement the theoretical analysis made by Sasselov and his colleagues, adding to the interest of such calculations. They first showed that there are two main kinds of super-Earth: those with a lot of water, and those with much less. The first kind would have formed quite a long way out from the parent star, where they would pick up large amounts of ice. The second kind would have formed further in, and be relatively dry. Both kinds would acquire a large iron core as the denser parts of their molten material sank towards the centre, and a silicate mantle as the lighter materials rose. The water-rich super-Earths would have very deep oceans above the mantle; the drier ones would have thin oceans, or none.

Because the pressure at the centre of a large super-Earth is higher than it is for our own world, the iron core will solidify faster. This probably implies that such a planet will have little or no magnetic field, and that might be bad for the occurrence of life, because magnetic fields shield the surface from radiation. However, so do deep oceans, and in any case we don't know how necessary a magnetic field really is. Some Earthly bacteria are radiation-resistant, for example.

The interior of a large super-Earth should contain the radioactive elements uranium and thorium, which generate most of the heat that keeps our own planet's core molten. Because these elements occur in much the same proportions throughout the Galaxy, the large super-Earth would have more of them than our own world does, and its core would be considerably hotter. The extra heat would cause convection in the mantle to be more vigorous, and this in turn would drive the movement of large plates at the boundary of the rock, much as it does on Earth. It turns out that these plates would be thinner than they are on Earth, because they move more rapidly and so have less time to thicken up by cooling. They would be easier to deform, except that the planet's greater gravity exerts more pressure on fault lines, so the plates don't slide as easily as they do here. These two effects tend to cancel out, so overall the frictional resistance when the plates slide past one another is much the same, regardless of size.

In short: plate tectonics is likely to be *more* common on large super-Earths than it is on Earth-like terrestrial planets that are similar in size to our own world. It also happens faster, which means that the cycle of subduction and volcanic activity that tends to keep the carbon dioxide concentration fairly stable would, if anything, work better. So a super-Earth that is considerably larger than our own world would probably have a more stable climate than ours, on geological timescales, making it easier for complex life to evolve.

This analysis completely changes the 'rare Earth' picture. Terrestrial planets, roughly the same size as our own, should occur fairly often, but comparatively speaking they are probably fairly rare. But the likely number of super-Earths in the Galaxy is far greater than the number of terrestrial planets, so the prospects for life are much better than they would appear if we were to focus solely on terrestrial planets. It also casts serious doubt on Goldilocks arguments, because it turns out that the Earth, far from being 'just right' for plate tectonics to arise, is very close to the lower extreme of the range of sizes for which such effects can happen. If the Earth were slightly smaller, it would not have plate tectonics, and that might have caused it not to evolve complex life.

The ideal Earth-like planet, it seems on this analysis, is considerably larger than the Earth. We just scraped into the acceptable range.

There is a general message in this work, and it is one that needs to be far more widely appreciated. The way to understand how likely alien life might be is not to focus on conditions that are virtually identical to those found on this world, and then argue – typically confusing sufficiency with necessity – that only those conditions are suitable for life. What really matters is just how *different* a planet can be from ours, and still support its own form of life, adapted to its prevailing conditions by evolution.

How diverse can living creatures, and their worlds, be? You won't find out if you start by assuming they all have to be just like us.

. . . .

Perhaps alien life has already been discovered.

In 1997 NASA launched the Cassini–Huygens spacecraft, a mission to Saturn. Seven years later the craft reached its destination. The Huygens probe landed on one of Saturn's moons, Titan. The Cassini spacecraft went into orbit round the planet. One of the early discoveries – dramatic, though to some extent expected – was that Titan has lakes. Because of the deep cold at that distance from the Sun, the lakes are not of water, but liquid methane and ethane.

Now some scientists are wondering whether Cassini has found signs of an exotic kind of life. This is one possible explanation for the strange behaviour of two gases on Titan: hydrogen and acetylene. There ought to be quite a lot of hydrogen, spread fairly uniformly in the moon's atmosphere. Darrell Strobel, working at Johns Hopkins University, has discovered that the hydrogen streams downwards through the atmosphere and disappears near the surface. Astronomers had expected acetylene to be fairly common too, produced by simple chemical reactions in Titan's atmosphere and deposited on the surface.

But there isn't any.

In 2005 Chris MacKay, a planetary scientist at NASA, realised that hypothetical methane-based microbial life would be very likely to get its energy by reacting hydrogen with acetylene, in the same way that most Earthly life reacts oxygen with molecules that contain carbon. The new observations are consistent with this kind of life inhabiting Titan's surface, and using up all of the missing hydrogen and acetylene.

Of course this doesn't come close to a proof that Titan harbours such exotic life forms, and Mark Allen (also at NASA) has suggested that non-living processes are a more likely explanation. Cosmic rays could convert acetylene to more complex substances, for instance, when they collide with its molecules. But it does illustrate the value of not assuming that life everywhere must be very much like life here. By doing so, we could have stood a small but significant chance of missing an alien life form in our own backyard.

Watch this space.

19 The Sixth Revolution

· ·

Mathematics has played a central role in the physical sciences for hundreds of years. In 1623, in *The Assayer*, Galileo wrote:

> Philosophy is written in this grand book, the universe, which stands continually open to our gaze. But the book cannot be understood unless one first learns to comprehend the language and read the characters in which it is written. It is written in the language of mathematics, and its characters are triangles, circles, and other geometric figures, without which it is humanly impossible to understand a single word of it; without these one is wandering in a dark labyrinth.

His words were prophetic. By the seventeenth century, mathematics had become a major driving force behind dramatic advances in the physical sciences, and today mathematics and physics (along with astronomy, chemistry, engineering and related areas) have become inseparable.

Until fairly recently, however, mathematics played a much smaller role in the development of the biological sciences. One reason is the old joke about a farmer who hires some mathematicians to help him improve his milk yield. When they present him with their report, he opens it, only to read the opening sentence: 'Consider a spherical cow.' Galileo's language of triangles and circles seems far removed from the organic forms of the living world. You don't find a cow in Euclid.

This story is amusing, and holds a lesson for wannabee biomathematicians. But it also reveals a misunderstanding about mathematical models. They don't have to be an exact representation of reality to be useful. In fact, making them less realistic generally makes them more useful, as long as they still provide useful insights. A model that is as complex as the process or thing it represents is likely to be too complex to be useful. A simple model is easier to work with. So a spherical cow is useless if you want it to give birth to a calf, but it might be a useful approximation if you're wondering about the spread of some bovine skin disease.

A good model must, of course, be sufficiently realistic that it doesn't leave out anything of vital importance. If you model a rabbit population using immortal rabbits, you will observe a population explosion that has little to do with reality. But even then, your model may capture how a small population grows before it hits environmental limits – so don't dismiss it too readily. What counts is what the model predicts, not what it leaves out.

Part of the art of biomathematics is the selection of useful models. Another part is taking the biology seriously, and not missing out anything crucial. A third is to pay attention to the problems that biologists want to solve. But sometimes it is also necessary to take a step back, try out a new mathematical idea in a simple but unrealistic setting, and see where that leads. There is another old joke, about a drunk searching under a lamppost for his keys. 'Did you drop them here?' 'No, but this is the only place where there's enough light to look.' It is not widely appreciated that the joke's original context, in *Computer Power and Human Reason* by Joseph Weizenbaum, was an analogy with science. The point was that in science you *have* to search under the lamppost, or you'll never find anything. Maybe, just maybe, you'll find a torch, even if the keys are somewhere along the road in the gutter. Several of the topics in *Mathematics of Life* started out as wild oversimplifications, the best that could be done at the time, but eventually turned out to be really informative about biology. It's important not to strangle a good idea at birth.

Looking back on the story of how biology started to embrace mathematics, one thing stands out: it was doing so long before anyone noticed. Mendel's discoveries hinged on simple mathematical patterns in the numbers of plants with particular characters. Although the early development of the microscope was empirical, the mathematics of optics soon entered into the story, because you can't develop really good microscopes without it. One of the clues to the structure of DNA was Chargaff's rule, a striking but unexplained numerical relationship that couldn't be coincidence. Bragg's law for X-ray diffraction was also crucial, and much of what we know about the structure of biologically significant molecules depends on it. And although evolution did not acquire any mathematical expression until recently, Darwin was on the *Beagle* because, among other activities, the vessel was carrying out a chronometric survey – a mathematical technique for finding longitude.

My sixth revolution, then, is not revolutionary because no one ever used mathematics to solve a biological problem before. What is revolutionary is the breadth of the methods used, and the extent to which they are starting to set the agenda in some areas of biology. I doubt that mathematics will ever dominate biological thinking in the way it now does for physics, but its role is becoming essential. In the twenty-first century, biology makes use of mathematics in ways that no one would have dreamed of at the start of the twentieth. By the time we get to the twenty-second century, mathematics and biology will have changed each other out of all recognition, just as mathematics and physics did in the nineteenth and twentieth centuries.

In Darwin's day, geology, not mathematics, was vital to the nascent theory of evolution. In the 1960s, chemistry became an essential foundation for cell biology. Then computer science joined in, with the advent of bioinformatics. Now physics and mathematics are entering the fray. And it's not just biology that is changing in this way: so are all the other branches of science. Conventional borders in science are breaking down. You can no longer study biology as if the rest of science didn't exist.

Instead of isolated clusters of scientists, obsessed with their own narrow speciality, today's scientific frontiers increasingly require teams of people with diverse, complementary interests. Science is changing from a collection of villages to a worldwide community. And if the story of mathematical biology shows anything, it is that interconnected communities can achieve things that are impossible for their individual members.

Welcome to the global ecosystem of tomorrow's science.

Notes

Chapter 1: Mathematics and Biology

1 More precisely, the common house cat is *Felis sylvestris catus*, but its binomial name is *Felis catus*.

2 J.D. Watson and F.H. Crick, 'Molecular structure of nucleic acids: a structure for deoxyribose nucleic acid', *Nature* **171** (1953) 737–738.

3 The date depends on what you count as 'completion'. A draft sequence was published in 2000, a so-called 'complete' draft in 2003. The sequence for the final chromosome, chromosome 1, was published in May 2006 in *Nature*. Some gaps remain, so it is arguable that the task is not yet finished. There are several thousand known gaps, inconsistencies and errors, currently being tidied up by a dedicated team of biologists.

Chapter 2: Creatures Small and Smaller

1 For an animation, see www.cellimagelibrary.org/images/8082

Chapter 3: Long List of Life

1 Most taxonomists consider *Cyanistes* to be a subgenus of the genus *Parus*, but the British Ornithologists' Union considers it to be a distinct genus, on the basis of DNA sequencing (specifically, the mitochondrial DNA sequence of cytochrome B) which shows that these birds differ significantly from the other tits. As regards finer distinctions, *C. caeruleus* subdivides into at least nine subspecies.

Chapter 4: Florally Finding Fibonacci

1 H. Vogel, 'A better way to construct the sunflower head', *Mathematical Biosciences* **44** (1979) 179–189.

2 S. Douady and Y. Couder, 'Phyllotaxis as a self-organised growth

process', in *Growth Patterns in Physical Sciences and Biology* (ed. J.-M. Garcia-Ruiz *et al.*), Plenum Press, New York (1993) 341–351.

3   L.S. Levitov, 'Phyllotaxis of flux lattices in layered superconductors', *Physics Review Letters* **66** (1991) 224–227; M. Kunz, 'Some analytical results about two physical models of phyllotaxis', *Communications in Mathematical Physics* **169** (1995) 261–295.

4   The species is *Echinocactus grusonii inermis*. See www.maths.surrey.ac. uk/hosted-sites/R.Knott/Fibonacci/fibnat.html#nonfib

5   G.W. Ryan, J.L. Rouse and L.A. Bursill, 'Quantitative analysis of sunflower seed packing', *Journal of Theoretical Biology* **147** (1991) 303–328.

6   P.D. Shipman and A.C. Newell, 'Phyllotactic patterns on plants', *Physics Review Letters* **92** (2004) 168102.

7   A.C. Newell, Zhiying Sun and P.D. Shipman, 'Phyllotaxis and patterns on plants', preprint, University of Arizona 2009.

Chapter 5: The Origin of Species

1   'Presidential Address', *Proceedings of the Linnaean Society*, 24 May (1859) viii.

2   F. Darwin (ed.), *The Foundations of The Origin of Species. Two essays written in 1842 and 1844.* Cambridge University Press, Cambridge (1909).

3   At first sight it is difficult for a modern mind to understand how anyone could arrive at such a specific date. The Book of Genesis, for instance, does not tell us how long Adam and Eve lived in the Garden of Eden before being expelled. But Ussher's deductions from genealogical records in the Old Testament convinced him that the date of creation was precisely 4,000 years before the birth of Christ. If Ussher could date the nativity accurately, he would automatically date the creation. At that time the consensus among theologians was that Jesus was born in 4 BC, hence 4004 BC for the Creation. Ussher could date other Biblical events as well: Noah's flood, he found, occurred in 2348 BC.

4   A survey by Gallup in 2004 indicated that about 45% of Americans accept both a 10,000-year-old Earth and the divine origin of the planet, 38% assigned the Earth's origin to God but preferred a timescale of millions of years, and 13% believed that it took millions of years and God played no part in the process. In a 1997 Gallup poll of Americans with science degrees, only 5% thought that the Earth was less than 10,000 years old. Another 40% accepted divine creation, but

placed it millions of years in the past. The remaining 55% believed that the Earth was extremely ancient, and that God played no role in the evolution of humans. Among those earning less than $20,000 per year, the corresponding figures were 59%, 28% and 6.5%; among those earning more than $50,000 per year they were 29%, 50% and 17%.

5   The poem was written in 1849, twenty years before the *Origin* appeared. But it was influenced by the 1844 book *Vestiges of the Natural History of Creation*, published anonymously by Robert Chambers. This book described the transmutation of species, the evolution of stars and other speculative scientific theories, and softened up public opinion for later ideas about evolution. It found favour with radicals, but once its implications had sunk in, it was denounced by the establishment for its alleged materialism.

6   Think of football. Ignoring draws, one team must win and one team must lose in each match. But winning is not purely random. Teams that have greater skill with the ball tend to win more matches. If we define 'skill' tautologously, in terms of which team wins, the above statement will still be true. However, that's the start of understanding, not the end. On closer examination we can discover which abilities with the ball, or strategy, or strength, or passion or 'belief' make teams more likely to win than others. If we could kill off losing teams, and clone winning ones, along with their skills, the standard of play would generally improve.

     Louis Amaral used network methods to analyse the skills of teams in the 2008 UEFA European Football Championship, assigning points for precision in passing, shots at goal, and so on. These data, derived from video footage of the games, were used to assign a skill level to each team. This ranking closely matched the actual results in the tournament. See J. Duch, J.S. Waitzman and L.A.N. Amaral, 'Quantifying the performance of individual players in a team activity', *PLoS ONE* 2010 5(6): e10937. doi:10.1371/journal.pone.0010937.

7   Two small herbivorous dinosaurs are happily eating plants when they spot an approaching velociraptor. (It used to be tyrannosaur, but we are in the post-*Jurassic Park* era now.) One of them immediately starts to run. 'There's no point in running away,' says the other. 'You can't outrun a velociraptor.' The first one turns and shouts back over its shoulder, 'No, but I can run faster than *you!*'

8   Actually, this is a simplification. Some regions of the genome are more likely to change than others, for instance.

9   Theodosius Dobzhansky, 'Nothing in biology makes sense except in the light of evolution', *American Biology Teacher* **35** (1973) 125–129.

10   In 1841 Richard Owen, a leading palaeontologist, found an incomplete fossil that he thought was a hyrax (because of its teeth) and assigned it to a new genus, *Hyracotherium*. In 1876 Othniel Marsh, Owen's rival, discovered a complete skeleton, obviously horse-like, and assigned it to another new genus, *Eohippus* (dawn horse). Later it became clear that the two fossils belonged to the same genus, and by the rules of taxonomy the name that was the first to be published won. So the evocative 'dawn horse' was lost, and a scientific misconception was preserved.

Chapter 6: In a Monastery Garden

1   Perhaps too well: reanalysis of Mendel's data suggests that the fit is better than we should expect statistically. Perhaps there was some subconscious massaging of the data in ambiguous cases. See R.A. Fisher, 'Has Mendel's work been rediscovered?', *Annals of Science* **1** (1936) 115–137.

2   This convention is not obvious. Usually the symbols are standardised so that in AB the factor A comes from the father and B from the mother, so AB and BA are potentially distinguishable. Mendel's experiments were what suggested that AB=BA.

Chapter 7: The Molecule of Life

1   Originally deoxyribose nucleic acid. See the quote from Crick and Watson on p. 6.

2   This may seem a chicken-and-egg situation: you need DNA to specify the enzymes, and you need the enzymes to copy DNA. Like all such puzzles, the answer is presumably that this feedback loop had simpler origins, without this recursive structure.

3   Mitochondria of animals or microorganisms (but not plants) use UGA (U=uracil) to encode tryptophan rather than STOP. When translated by the cell's molecular machinery, synthesis stops where tryptophan should have been inserted. In addition, most animal mitochondria use AUA for methionine, not isoleucine; vertebrate mitochondria use AGA and AGG as STOP, and yeast mitochondria assign all triplets beginning with CU to threonine instead of leucine.

4   Richard Dawkins, *The Selfish Gene*, Oxford University Press, Oxford (1989).

5   Meaning 'we don't understand what this bit does', and typically confused with 'this bit does nothing'.

6   A bizarre misconception sometimes inspired by this term is that your DNA makes *you* selfish.

7   Jack Cohen and Ian Stewart, *The Collapse of Chaos*, Viking, New York (1994).

8   John S. Mattick, 'The hidden genetic program of complex organisms', *Scientific American*, October 2004, **291**(4) 60–67.

Chapter 8: The Book of Life

1   US Governmental agencies have a long track record of funding research that seems outside their natural remit. The project was handled by a special subcommittee of the DoE's Health and Environmental Research Advisory Committee.

2   In 2010 a US court ruled that nine patents filed by a company called Myriad, related to the so-called breast cancer genes *BRCA1* and *BRCA2*, are invalid. As I write, the decision is being appealed.

3   At least not yet. But nanotechnology opens up the possibility of pulling a strand of DNA through a device that can read off the bases by exploiting subtle differences in their electronic properties. Some progress has already been made; the latest is to make a small hole in a layer of graphene, which is a honeycomb of carbon atoms just one atom thick.

4   T. Radford, '"Gay gene" theory fails blood test', *The Guardian* (23 April 1999) 11.

5   D.H. Hamer, S. Hu, V.L. Magnusson, N. Hu and A.M. Pattatucci, 'A linkage between DNA markers on the X chromosome and male sexual orientation', *Science* **261** (1993) 321–327.

6   G. Rice, C. Anderson, N. Risch and G. Ebers, 'Male homosexuality: absence of linkage to microsatellite markers at Xq28', *Science* **284** (1999) 665–667.

Chapter 9: Taxonomist, Taxonomist, Spare that Tree

1   K. Ochiai, T. Yamanaka, K. Kimura and O. Sawada. 'Inheritance of drug resistance (and its transfer) between *Shigella* strains and between *Shigella* and *E. coli* strains' [in Japanese], *Hihon Iji Shimpor* **1861** (1959) 34.

2   D.L. Theobald, 'A formal test of the theory of universal common ancestry', *Nature* **466** (2010) 219–222.

Chapter 10: Virus from the Fourth Dimension

1   Euclid put geometry on a systematic basis, but he seems not to have created much new geometry of his own. Other Greek geometers, such as Apollonius, Eudoxus and Archimedes, are generally considered to have been more creative as mathematicians.

2   The cube can also be called a hexahedron, keeping the names consistent, but nobody does that any more.

3   For example, Archimedes' principle remains fundamental in the design of ships, because it governs whether they will float, and if so, how stable they will be. And the law of the lever is built into computer software for designing buildings, cars and bridges.

4   Ian Stewart, *Why Beauty is Truth*, Basic Books, New York (2007).

5   D.L.D. Caspar and A. Klug, 'Physical principles in the construction of regular viruses', *Cold Spring Harbor Symposia on Quantitative Biology* 27, Cold Spring Harbor Laboratory, New York (1962) 1–24.

6   N.G. Wrigley, 'An electron microscope study of the structure of *Sericesthis* iridescent virus', *Journal of General Virology* 5 (1969) 123–134; N.G. Wrigley, 'An electron microscope study of the structure of *Tipula* iridescent virus', *Journal of General Virology* 5 (1970) 169–173.

7   R.C. Liddington, Y. Yan, J. Moulai, R. Sahli, T.L. Benjamin and S.C. Harrison, 'Structure of simian virus 40 at 3.8-Å resolution', *Nature* 354 (1991) 278–284.

8   R. Twarock, 'A mathematical physicist's approach to the structure and assembly of viruses', *Philosophical Transactions of the Royal Society of London A* 364 (2006) 3357–3374.

9   R. Twarock, 'A tiling approach to virus capsid assembly explaining a structural puzzle in virology', *Journal of Theoretical Biology* 226(4) (2004) 477–482.

10 If you're wondering how Donald is extracted from H.S.M., the M stands for MacDonald.

Chapter 11: Hidden Wiring

1   S. Herculano-Houzel, B. Mota and R. Lent, 'Cellular scaling rules for rodent brains', *Proceedings of the National Academy of Sciences* 103 (2006) 12138–12143; S. Herculano-Houzel, 'The human brain in numbers: a linearly scaled-up primate brain', *Frontiers in Human Neuroscience* 3 (2009) article 31.

2   A. Hodgkin and A. Huxley, 'A quantitative description of membrane

current and its application to conduction and excitation in nerve', *Journal of Physiology* **117** (1952) 500–544.

3   The Hodgkin–Huxley equations take the form

$$I = C\frac{dV}{dt} + g_K n^4 (V - V_K) + g_{Na} m^3 h(V - V_{Na}) + g_L (V - V_L)$$

where I=membrane current, C=membrane capacitance, V=voltage, V_K, V_{Na} and V_L are constants related to the potassium, sodium and other ion channels, and m, n and h are determined by three differential equations based on observed data.

4   The FitzHugh–Nagumo equations (with no applied current) are:

$$\frac{dv}{dt} = v(a - v)(v - 1) - w$$
$$\frac{dw}{dt} = bv - gw$$

where v is a dimensionless form of the voltage V, and w combines the roles of m, n and h in the Hodgkin–Huxley equations into a single variable.

5   M. Golubitsky, I. Stewart, P.-L.Buono and J.J. Collins, 'Symmetry in locomotor central pattern generators and animal gaits', *Nature* **401** (1999) 693–695.

6   C.A. Pinto and M. Golubitsky, 'Central pattern generators for bipedal locomotion', *Journal of Mathematical Biology* **53** (2006) 474–489.

7   R.L. Calabrese and E. Peterson, 'Neural control of heartbeat in the leech *Hirudo medicinalis*' in *Neural Origin of Rhythmic Movements* (ed. A. Roberts and B. Roberts), *Symposium of the Society for Experimental Biology* **37** (1983) 195–221; E. De Schutter, T.W. Simon, J.D. Angstadt and R.L. Calabrese, 'Modeling a neuronal oscillator that paces heartbeat in the medicinal leech', *American Zoologist* **33** (1993) 16–28; R.L. Calabrese, F. Nadim and Ø.H. Olsen, 'Heartbeat control in the medicinal leech: a model system for understanding the origin, coordination, and modulation of rhythmic motor patterns', *Journal of Neurobiology* **27** (1995) 390–402; W.B. Kristan Jr, R.L. Calabrese and W.O. Friesen, 'Neuronal control of leech behavior', *Progress in Neurobiology* **76** (2005) 279–327.

8   P.-L. Buono and A. Palacios, 'A mathematical model of motorneuron dynamics in the heartbeat of the leech', *Physica D* **188** (2004) 292–313.

9   H.R. Wilson and J.D. Cowan, 'Excitatory and inhibitory interactions in localized populations of model neurons', *Biophysical Journal* **12** (1972) 1–24.

10 P.C. Bressloff, J.D. Cowan, M. Golubitsky and P.J. Thomas, 'Scalar and pseudoscalar bifurcations motivated by pattern formation on the visual cortex', *Nonlinearity* **14** (2001) 739–775; P.C. Bressloff, J.D. Cowan, M. Golubitsky, P.J. Thomas and M.C. Wiener, 'Geometric visual hallucinations, Euclidean symmetry, and the functional architecture of striate cortex', *Philosophical Transactions of the Royal Society of London B* **356** (2001) 299–330; P.C. Bressloff, J.D. Cowan, M. Golubitsky, P.J. Thomas and M.C. Wiener, 'What geometric visual hallucinations tell us about the visual cortex', *Neural Computation* **14** (2002) 473–491.

11 J.W. Zweck and L.R. Williams, 'Euclidean group invariant computation of stochastic completion fields using shiftable–twistable functions', *Journal of Mathematical Imaging and Vision* **21** (2004) 135–154.

Chapter 12: Knots and Folds

1 S.A. Wasserman, J.M. Dungan and N.R. Cozzarelli, 'Discovery of a predicted DNA knot substantiates a model for site-specific recombination', *Science* **229** (1985) 171–174; D. Sumners, 'Lifting the curtain: using topology to probe the hidden action of enzymes', *Notices of the American Mathematical Society* **42** (1995) 528–537.

2 Animations showing how haemoglobin changes shape when binding to oxygen, or releasing it, can be found at en.wikipedia.org/wiki/Hemoglobin#Binding_for_ligands_other_than_oxygen.

3 I'm simplifying a complicated tale by talking of haemoglobin as if it were unique. Actually, many variants of the specific form of haemoglobin shown in Figure 51 exist in nature. They are all fairly similar and probably have a common evolutionary origin. The point is stronger: in principle, vast numbers of radically different molecules could also transport oxygen. 'The right shape' here doesn't mean 'the only shape': it means any shape that will do the job. Lots will; far more won't.

4 C. Levinthal, 'How to fold graciously', in *Mössbauer Spectroscopy in Biological Systems* (ed. J.T.P. DeBrunner and E. Munck), University of Illinois Press, Illinois (1969) 22–24.

5 C.M. Dobson, 'Protein folding and misfolding', *Nature* **426** (2003) 884–890.

6 S. Cooper, F. Khatib, A. Treuille, J. Barbero, J. Lee, M. Beenen, A. Leaver-Fay, D. Baker, Z. Popović and Foldit players, 'Predicting protein structures with a multiplayer online game', *Nature* **466** (2010) 756–760.

7 You can try Foldit for yourself at http://fold.it/portal/

Chapter 13: Spots and Stripes

1 A.M. Turing, 'The chemical basis of morphogenesis', *Philosophical Transactions of the Royal Society of London B* **237** (1952) 37–72.
2 J. Murray, *Mathematical Biology*, Springer, Berlin (1989).
3 S. Kondo and R. Asai, 'A reaction–diffusion wave on the skin of the marine angelfish *Pomacanthus*', *Nature* **376** (1995) 765–768.
4 The precise statement of 'a lot of symmetry' is called the maximal isotropy subgroup conjecture. This plausible conjecture was eventually proved false: see M.J. Field and R.W. Richardson, 'Symmetry breaking and the maximal isotropy subgroup conjecture for reflection groups', *Archive for Rational Mechanics and Analysis* **105** (1989) 61–94.
5 H. Meinhardt, 'Models of segmentation', in *Somites in Developing Embryos* (ed. R. Bellairs *et al.*), Nato ASI Series A 118, Plenum Press, New York (1986) 179–189.
6 P. Eggenberger Hotz, 'Combining development processes and their physics in an artificial evolutionary system to evolve shapes', in *On Growth, Form, and Computers* (ed. S. Kumar and P.J. Bentley), Elsevier, San Diego CA (2003) 302–318.

Chapter 14: Lizard Games

1 J. Maynard Smith, *Evolution and the Theory of Games*, Cambridge University Press, Cambridge (1982) 16.
2 M. Pigliucci, 'Species as family resemblance concepts: the (dis-)solution of the species problem?', *BioEssays* **25** (2003) 596–602.
3 N. Knowlton, 'Sibling species in the sea', *Annual Review of Ecology and Systematics* **24** (1993) 189–216.
4 This is not straightforward, and some aspects are controversial. Some regions of the genome are conserved by natural selection: even if mutations happen, they don't survive.
5 L.M. Mathews and A. Anker, 'Molecular phylogeny reveals extensive ancient and ongoing radiations in a snapping shrimp species complex (Crustacea, Alpheidae, *Alpheus armillatus*)', *Molecular Phylogenetics and Evolution* **50** (2009) 268–281.
6 G. Vogel, 'African elephant species splits in two', *Science* **293** (2001) 1414.
7 H. Sayama, L. Kaufman and Y. Bar-Yam, 'Spontaneous pattern formation and genetic diversity in habitats with irregular geographical features', *Conservation Biology* **17** (2003) 893; M.A.M. de Aguiar, M. Baranger, Y. Bar-Yam and H. Sayama, 'Robustness of spontaneous

pattern formation in spatially distributed genetic populations',
Brazilian Journal of Physics **33** (2003) 514–520.

8   A.S. Kondrashov and F.A. Kondrashov, 'Interactions among
quantitative traits in the course of sympatric speciation', *Nature* **400**
(1999) 351–354.

9   U. Dieckmann and M. Doebeli, 'On the origin of species by sympatric
speciation', *Nature* **400** (1999) 354–357.

10 The species of Darwin's finch are:
Large cactus-finch, *Geospiza conirostris*
Sharp-beaked ground-finch, *Geospiza difficilis*
Vampire finch, *Geospiza difficilis septentrionalis* [subspecies]
Medium ground-finch, *Geospiza fortis*
Small ground-finch, *Geospiza fuliginosa*
Large ground-finch, *Geospiza magnirostris*
Darwin's large ground-finch, *Geospiza magnirostris magnirostris* [possibly
extinct subspecies]
Common cactus-finch, *Geospiza scandens*
Vegetarian finch, *Camarhynchus crassirostris*
Large tree-finch, *Camarhynchus psittacula*
Medium tree-finch, *Camarhynchus pauper*
Small tree-finch, *Camarhynchus parvulus*
Woodpecker finch, *Camarhynchus pallidus*
Mangrove finch, *Camarhynchus heliobates*
Warbler finch, *Certhidea olivacea*

11 That term was made popular by the second book, but it originated
with Percy Lowe in 1936.

12 More precisely, the mean phenotype changes continuously, even
though the two new phenotypes appear through jumps. If we consider
how phenotypes deviate from their average values, the word
'unchanged' applies, and this is what the mathematical models
actually study. J. Cohen and I. Stewart, 'Polymorphism viewed as
phenotypic symmetry breaking', in *Nonlinear Phenomena in Biological
and Physical Sciences* (ed. S.K. Malik *et al.*), Indian National Science
Academy, New Delhi (2000) 1–63; I. Stewart, T. Elmhirst and J. Cohen,
'Symmetry breaking as an origin of species', in *Bifurcations, Symmetry,
and Patterns* (ed. J. Buescu *et al.*), Birkhäuser, Basel (2003) 3–54.

Chapter 15: Networking Opportunities

1   A. Tero, S. Takagi, T. Saigusa, K. Ito, D.P. Bebber, M.D. Fricker, K. Yumiki, R. Kobayashi and T. Nakagaki, 'Rules for biologically inspired adaptive network design', *Science* **327** (2010) 439–442.

2   Translating his symbolic terminology into features of the diagram, what matters is how many lines meet at a given dot. Suppose, for example, that a closed path exists. Then whenever the path runs into a dot, it also departs from that dot. So the number of lines meeting any given dot must be even. This disposes of the Königsberg bridges, because the diagram has three dots where three lines meet, and one dot where five lines meet. Since the number of lines is odd in these cases, the puzzle cannot be solved with a closed path. Open paths have two distinct ends, and at each of these the number of lines running into the dot is odd. Everywhere else, it is even. So now there must be exactly two dots that meet an odd number of lines: one at each end of the path. Since the Königsberg diagram has four dots meeting an odd number of lines, there is no open path either.

Euler proved that these conditions are also sufficient for a path of the appropriate kind to exist provided the diagram is connected (any two dots are joined by some path). Euler devotes several pages to a symbolic proof; in diagrammatic form it can be made virtually obvious.

3   Y. Kuramoto, *Chemical Oscillations, Waves, and Turbulence*, Springer, New York (1984).

4   J.R. Collier, N.A.M. Monk, P.K. Maini and J.H. Lewis, 'Pattern formation by lateral inhibition with feedback: a mathematical model of Delta–Notch intercellular signalling', *Journal of Theoretical Biology* **183** (1996) 429–446.

Chapter 16: The Paradox of the Plankton

1   Specifically, it is the integer closest to $\frac{1}{10}(5 + \sqrt{5})\,\phi^n$, where ϕ is the golden number $\frac{1}{2}(1 + \sqrt{5}) \sim 1.618034$.

2   The equation is

$$\frac{dN}{dt} = rN\left(1 - \frac{N}{K}\right)$$

where r and N are constants; here we can interpret r as the unconstrained growth rate, and K is the maximum population size. The actual growth rate, at population N, is $r(1 - N/K)$, which depends on N: such a growth rate is said to be density-dependent.

With initial conditions $N(0)=N_0$, the logistic equation can be solved explicitly by standard methods, and the solution is

$$N(t) = \frac{KN_0 e^{rt}}{K + N_0(e^{rt} - 1)}$$

where N_0 is the initial population.

3   R.M. May, 'Simple mathematical models with very complicated dynamics', *Nature* **261** (1976) 459–467.

4   The trick is to scale X_t to bX_t/a.

5   The model is:

$L_{t+1}=bA_t\exp(-c_{ca}A_t-c_{cl}L_t)$

$P_{t+1}=L_t(1-\mu_1)$

$A_{t+1}=P_t\exp(-c_{pa}A_t)+A_t(1-\mu_a)$

where L_t is the number of feeding larvae, P_t is the number of non-feeding larvae, pupae and newly emerged adults, and A_t is the number of adults, all at time t. The other symbols are parameters.

6   *Period-2*: R.F. Costantino, J.M. Cushing, B. Dennis and R.A. Desharnais, 'Experimentally induced transitions in the dynamic behavior of insect populations', *Nature* **375** (1995) 227–230; *Chaos*: R.F. Costantino, R.A. Desharnais, J.M. Cushing and B. Dennis, 'Chaotic dynamics in an insect population', *Science* **275** (1997) 389–391.

7   J. Huisman and F.J. Weissing, 'Biodiversity of plankton by species oscillations and chaos', *Nature* **402** (1999) 407–410.

8   E. Benincà, J. Huisman, R. Heerkloss, K.D. Jöhnk, P. Branco, E.H. Van Nes, M. Scheffer and S.P. Ellner, 'Chaos in a long-term experiment with a plankton community', *Nature* **451** (2008) 822–825.

9   M.J. Keeling, 'Models of foot-and-mouth disease', *Proceedings of the Royal Society of London B* **272** (2005) 1195–1202.

Chapter 17: What is Life?

1   There was a forerunner, called Project Ozma, in 1960.

2   For a full specification see en.wikipedia.org/wiki/ Von_Neumann_cellular_automata

3   J. Von Neumann and A.W. Burks, *Theory of Self-Reproducing Automata*, University of Illinois Press, Chicago (1966).

4   See e.g. www.ibiblio.org/lifepatterns/

5   E.R. Berlekamp, J.H. Conway and R.K. Guy, *Winning Ways* volume 2, Academic Press, London (1982).

6   M. Cook, 'Universality in elementary cellular automata', *Complex Systems* **15** (2004) 1–40.

7   C.G. Langton, 'Artificial life', in *Artificial Life* (ed. C.G. Langton), Addison-Wesley, Reading MA (1989), 1.

8   www.darwinbots.com/WikiManual/index.php/Main_Page

9   'Smallest' here is usually interpreted as 'take any more DNA away and it won't work'. Such a genome need not be unique, because there might be several different ways to cut a genome down to size, each working for different reasons.

Chapter 18: Is Anybody Out There?

1   K. Thomas-Keprta, S. Clemett, D. McKay, E. Gibson and S. Wentworth, 'Origin of magnetite nanocrystals in Martian meteorite ALH 84001', *Geochimica et Cosmochimica Acta* **73** (2009) 6631–6677.

2   D. Brownlee and P.D. Ward, *Rare Earth*, Copernicus, New York (2000).

3   Both words are Latin–Greek hybrids, which used to be unacceptable but are now so common that few object to them. 'Television' is a case in point.

4   Let there be n planets, with n large, and let $p = 1/n$. The binomial distribution tells us that the probability of getting exactly one planet with intelligent life is $np(1-p)^{n-1}$, which is very close to e^{-1} because $np=1$ and $(1-1/n)^{n-1}$ is close to e^{-1}. Here $e=2.718$ is the base of natural logarithms, so $e^{-1}=0.37$. The probability of no such planets is $(1-p)^n$, which is the same. The probability of two or more is $1-0.37-0.37=0.26$.

5   'Currently' may seem to contradict relativity, which says that no concept of simultaneity can be consistent for all inertial observers. That may be true, but we can adopt a privileged frame of reference – our own. The intelligent aliens can then be current *from our point of view*.

6   C.D.B. Bryan, *Close Encounters of the Fourth Kind*, Alfred A. Knopf, New York (1995).

7   G. Wächtershäuser, 'Groundworks for an evolutionary biochemistry: the iron–sulphur world', *Progress in Biophysics and Molecular Biology* **58** (1992) 85–201.

8   N. Lane, 'Genesis revisited', *New Scientist* **2772** (17 August 2010) 36–39.

9   O.U. Mason, T.Nakagawa, M. Rosner, J.D. Van Nostrand, J. Zhou, A. Maruyama, M.R. Fisk and S.J. Giovannoni, 'First investigation of the microbiology of the deepest layer of ocean crust', *PlosOne*, 5(11): e15399.doi:10.1371/journal.pone.0015399.

10 This was the number on 11 January 2011.

11 M.R. Swain, G. Vasisht and Giovanna Tinetti, 'The presence of

methane in the atmosphere of an extrasolar planet', *Nature* **452** (2008) 329–331.

12 J.L. Bean, E.M.-R. Kempton and D. Homeier, 'A ground-based transmission spectrum of the super-Earth exoplanet GJ 1214b', *Nature* **468** (2010) 669–672.

13 J. Laskar, F. Joutel and P. Robutel, 'Stabilization of the Earth's obliquity by the Moon', *Nature* **361** (1993) 615–617.

14 The Apollo Moon missions brought back rocks which show that the Moon's surface has much the same composition as the Earth's mantle. The Moon also has a large angular momentum. An attractive way to explain both these facts is a collision between the Earth and another large body, which splashed off a big chunk of mantle, forming the Moon. Simulations show that a Mars-sized body could have done this. This 'giant impact hypothesis' became established, and the body concerned was named Theia. But later, improved simulations show that a big chunk of Theia ends up on the Moon too. So now it is assumed that Theia had almost exactly the same composition as the Earth's mantle. This puts us back where we were at the start, but with an extra body involved. I call this 'losing the plot'.

15 I'm simplifying the discussion slightly: the usual formula also includes the planet's emissivity – its ability to emit radiation. I've set that to the value 1, which is typical. The formula is

$$T_\mathrm{p} = T_\mathrm{s}\sqrt{\frac{R}{2D}}\left(1 - \frac{\alpha}{\varepsilon}\right)^{1/4}$$

where T_p is the temperature of the planet, T_s is the temperature of the star, R is the radius of the star, D is the distance from the star to the planet, α is the planet's albedo and ε is the planet's emissivity.

For the Sun–Earth system, T_s=5,800 K, R=700,000 km, α=0.3, and the Earth's average emissivity in the infrared region of the spectrum (which is where most of the energy is radiated away) is approximately given by ε=1. Substituting these figures into the formula yields T_p=254 K.

16 An example is the exoplanet HD 209458b.

17 The metalloid magma-dwellers of Grumbatula VI used to think that Sol III (the Earth) was far too cold to support life, being well outside the habitable zone of the Sun in which the surfaces of planets are molten rock. The discovery of Earth's subterranean magma ocean, containing a million times as much magma as all the oceans of Grumbatula VI put together, was at first greeted with incredulity

because there was no conceivable heat source to keep it molten, but has now provoked a rethink among the more imaginative astroscholars.

18   M. Dellnitz, K. Padberg, M. Post and B. Thiere, 'Set oriented approximation of invariant manifolds: review of concepts for astrodynamical problems', in *New Trends in Astrodynamics and Applications III* (ed. E. Belbruno), AIP Conference Proceedings **886** (2007) 90–99.

19   D. Sasselov and D. Valencia, 'Planets we could call home', *Scientific American* **303**(2) (August 2010) 38–45.

Acknowledgements

The following figures are reproduced with the permission of the named copyright holders:

Figures 4, 5, 15, 16, 20, 25 (*right*), 27, 31, 34, 51, 70, 73, 75: Wikimedia Commons. Reproduced under the terms of the GNU Free Documentation License.

Figure 10 (*right*): James Murray.

Figure 25 (*left*): Kenneth J. M. MacLean. *A Geometric Analysis of the Platonic Solids and other Semi-regular Polyhedra*. LHP Press (2008).

Figure 25 (middle): Christian Schroeder.

Figure 26: Rochester Institute of Technology.

Figure 39: Ronald Calabrese.

Figures 41, 43, 44: Paul Bressloff. From P.C. Bressloff, J.D. Cowan, M. Golubitsky and P.J. Thomas, 'Scalar and pseudoscalar bifurcations motivated by pattern formation on the visual cortex', *Nonlinearity* **14** (2001) 739–775.

Figure 47: From S.A. Wasserman, J.M. Dungan and N.R. Cozzarelli, 'Discovery of a predicted DNA knot substantiates a model for site-specific recombination', *Science* **229** (1985) 171–174. Reprinted with permission from AAAS.

Figure 52: Christopher Dobson, 'Protein folding and misfolding', *Nature* **426** (2003) 884–890.

Figure 53 (*left*): Teresa Zuberbühler.

Figure 54 (*top*): Harry Swinney.

Figure 54 (*bottom*): Erik Rauch.

Figure 56: Shigeaki Kondo and Rihito Asai.

Figure 57 (*left*): ASTER.

Figures 57 (*right*), 74: NASA.

Figure 62: Michael Doebeli.

Figure 64: A.L. Barabasi and E. Bonabeau, 'Scale-free networks', *Scientific American* **288** (May 2003) 60.

Figure 65: Seiji Takagi.

Index

Note: 'n' following page numbers indicates notes.

A

acetylene and life on Titan 317

acquired characteristics, inheritance of 58, 59

action potentials (voltage spikes) 162, 163, 164

adaptations to environment, Lamarck's views 59

adaptive radiation 240

adenine (A) 6, 94, 95

thymine pairing with 96, 97

adhesion molecules, cell 211

African elephant 232

airway, food-crossing, evolution 295–6

albedo 307

Alexander polynomial 188

algebra, symmetry 204–5

ALH 84001 meteorite 289

alien life (astrobiology) 285–6, 289–316, 332–4n

astronomical features required for 303–9

discovery 315–16

images 297–8

visiting Earth 292–4

alife (artificial life) 285, 286–7

alleles 5

in cladogram construction 129

in human genome, different 121

see also genes; mutations

Allen, Mark 316

allopatric speciation 228, 228–9, 230, 231, 232, 233

Alzheimer's disease 193

amino acids

base triplet sequences coding for see genetic code

energy landscape for chain of 194

see also proteins

Amiskwia 134

amoeba 18

angelfish stripes 202–3

animal(s)

form/morphology 199–203, 209–12

markings/patterns 199–203

movement 165–9

animalcules 17

Anomalocaris 134

antibiotic resistance genes, transfer between species 135

apes, human evolution from 75–6

see also primates

aquatic animals, transition to land animals, to 75, 295–6

Arabidopsis shoot growth 46

archaea 22, 125, 137

Aristotle 57

Arnold, Vladimir 266

artificial life 285, 286–7

Asai, Rihito 202
Assayer, The 317
assortative mating 235, 236, 239
asteroids hitting Earth 304–5
astronomy 14, 16, 153
   and biology *see* alien life
ATG (STOP) sequence 99, 100
automata, cellular 282–5
automated DNA sequencers 114
auxins 53
Avery–MacLeod–McCarty
   experiment 94
Avida 287
axons 160, 161, 162, 163
   electrical pulses in *see* electrical
     activity
   excitability 163–4, 165

B

bacteria 22
   DNA, historical studies showing
     function 93–4
   extremophilic 301
   gene transfer between species 135
   sex 23
   synthetic 288
   viruses of (phages) 5, 94, 140, 157
'balance of nature' metaphor for
   ecosystems 270
Bar-Yam, Yaneer 233
barnacles, Darwin on 63–4
bases (DNA) 6, 92, 94–9
   parity and pairing 94–7
     Chargaff's rules 95–6, 319
   triplet sequences coding for
     amino acids *see* genetic code
Bauersfeld, Walther 139
Beagle voyage 61–2, 63, 243, 319

beaks, finches 239, 244
   Darwin's 63, 242–4, 245
     Lack's observations 243–4
bees, Mendel's studies 77–8
beetles
   flour, population dynamics 268–9
   horned, mating strategies 226–7
bias in game theory 216–17
bifurcation(s), evolutionary trees
   126
bifurcation diagram, phyllotaxis
   50–1, 51, 52
binary fission, prokaryotic cell
   reproduction by 84, 86
biochemistry, plant growth 53
bioinformatics 10–11
biomathematics/mathematical
   biology (general aspects) 7–12
   as 6th revolution in biology 7–9,
     317–19
   growth in 11
biomechanics *see* mechanics
biotechnology and genetic
   engineering 103–7
biped gait 169
birds
   cluster analysis 225
   Darwin's observations 61–2, 62–3
   see also blue tit; finches; gull
     species; owls
BirdSym 136
black-backed gulls 222
black bodies, planets as 306–7
blending theory of inheritance 78
Blind Watchmaker 117
blood as metaphor for inherited
   characteristics 78
blue tit, classification 33–4
body weight and brain weight
   158–9

bots (artificial life) 287
Boveri, Theodor 25, 85
bovine spongiform encephalopathy
    (BSE; mad cow disease) 193
Bragg, William Lawrence and
    William Henry, and Bragg's law
    92–3, 319
brain 159–61
  interspecies size comparisons
    158–9
  visual system 171–80
bran bug, population dynamics
    268–9
branching in evolutionary trees *see*
    species
break-point creation for DNA
    sequencing 113, 114, 115–16
breeding/mating
  artificial 3
    Mendel's studies 79–84
  assortative 235, 236, 239
  strategies
    horned beetles 226–7
    lizards 213–14, 215–16
  surviving to breeding age 215
  see also reproduction
Bryngelson, Joseph 193
BSE (mad cow disease) 193
buckminsterfullerene 139, 145
Buono, Luciano 167, 171
Burgess Shale 134
Burk, Arthur 284

C

cacti 52–3
  peyote 171
Caenorhabditis elegans 160
Calabrese, Ronald 169–71
calcium ion leakage, axons 162
Callisto 309

Cambrian explosion 134
Cambridge–Edinburgh model of
    foot-and-mouth disease
    epidemic 272
canine species, cladogram 128–9
capsid and capsomer geometry
    139–49, 153–7, 325n
carbon-based lifeforms, alternatives
    to 280, 299–300
carbon dioxide, Earth 307, 308, 312,
    314
Carruthers, Eleanor 85
Caspar–Klug theory 143, 144–6,
    154, 157
Cassini–Huygens spacecraft 315–16
cat
  evolutionary relationships
      between various species of
      131–2
  visual perception 173–4
Celera Genomics 112, 114, 116, 121
celestial mechanics 311
cells 19, 21–7
  development and differentiation
      211–12
    networks in 255–7
  division 18, 22, 84–7
    shoot tip 44
    see also meiosis; mitosis
  microscopic observation 1–2,
      20–1, 24, 25
cellular automata 282–5
central pattern generator (CPG)
    167, 168, 169, 170
centrioles 25
centrosomes 25
CFTR 122
chain-terminating method of DNA
    sequencing 113–14

chaos
  populations 266–71
  shell pattern 201
chaperonins 192
character displacement in Darwin's
    finches 244, 245
Chargaff's parity rules 95–6, 319
Chase, Martha 94
cheetahs, evolutionary relationships
    with other feline species 131–2
chicken, game of (hawk–dove game)
    218, 219
chloroplast evolution 87
chordates
  development 33
  evolution 134
chromosomes 5, 22, 25, 85–6, 88–90
  in meiosis 88
  in mitosis 86
  recombination *see* recombination
Chrysopa (lacewing) species 102,
    235, 235–6
cladistics and cladograms 128–33,
    136, 137
Clapham, Roy 259
classes (taxonomy) 2, 33
classification *see* taxonomy and
    classification
climate and plate tectonics 312–13
  super-Earths 314
closed loops
  in knot theory 184, 185, 186, 187
  in neural networks 171
  in tree of life 128, 136
cluster analysis 225–8
Cocos plate 241, 242
Cohen, Jack 103, 294, 295, 296
Collier, Joanne 255
Collins, Jim 166–7

colour, inheritance of
  flower 79–84
  hair 88–9
combinatorics in Mendel's studies
    81–4
comets hitting Earth 303–4, 304,
    310, 311, 312
common ancestor 74, 131, 137
  humans and apes 75
  universal 58, 137
competition
  interspecies 70–1, 258
    exclusion by 258, 269–70
  intraspecies 70–1, 221
    birds for food 239
    lizard mating 213
  resource 3, 65, 70, 258, 270
complexity
  life 278–9
  population growth dynamics 266,
    268, 269
Computer Power and Human Reason
    318
computers, general public's, protein
    folding studies 196–7
concentric circles, perception 175,
    186
continental drift *see* plate tectonics
continued fractions 44
continuity in topology 183
continuum model (layered neuronal
    networks) 177–8, 179
Conway, John Horton
  cellular automaton (Game of Life)
    284–5, 286, 287
  knot theory 189
Cook, Matthew 285
coral reefs, Darwin on 62, 63
core of planets 314
cork, microscopic structure 20–1

Cormis, Carlo 287
CoRoT-7b (exoplanet) 313
correlation studies in genetics
    119–20
cortex, visual 173–80
Couder, Yves 50, 51
Cowan, Jack 174, 177
    and Wilson–Cowan equation 175
Coxeter groups 156, 157
Cozzarelli, Nicholas 189, 191
Craig Venter Institute 288
creationism and evolution 60, 62,
    63, 64, 66, 67, 75, 76
Creutzfeldt–Jakob disease 193
Crick and Watson 5–6, 7, 96, 97,
    98, 101
cubic lattice 155
Curie, Pierre 204
Cushing, James 268–9
Cyanistes caeruleus 33, 34
cystic fibrosis, genetic basis 122
cytosine (C) 6, 94, 95
    guanine pairing with 96, 97
    methylated 108
cytoskeleton (cell skeleton) 24, 86

D

D'Armati, Salvino 16
Darwin, Charles 3–4, 56, 59–69,
    126, 136, 242–3, 319
    finches 61, 61–2, 242–4, 244–5,
    329n
    Origin of Species 3, 4, 68–9, 220,
    223
    see also neo-Darwinists
Darwin, Erasmus 57–8
Darwinbots 287
Dawkins, Richard 72, 103, 117
Delta 256, 257
dendrites 160, 161

deoxyribonucleic acid see DNA
Descent of Man 75
development 26–9
    cellular see cells
    chordates 33
    DNA and 181
    embryonic 26–9, 199
    networks in 255–7
Dieckmann–Doebeli model 235, 236
diffusion 200
digital messages in information
    theory 153
dimensions in space and viral
    structure 147–57
disease (predominantly infectious
    disease)
    endemic 259
    epidemics see epidemics
    epidemiological studies 259
    genes and risk of 118, 122
    transmission 254, 259
DNA (deoxyribonucleic acid) 5–7,
    91–110, 181, 185–90, 323–4n
    copying/replication 97–8, 185–6
    eukaryotic 23
    function 93–101
    'junk' 181
    life based on 277, 279, 299
        alternatives to/non-uniqueness
        of 279–80, 287, 299–300,
        301
    methylation 108
    mitochondrial 87, 100, 130, 323n
    prokaryotic 23
    sequencing data
        in cladogram construction 130
        in speciation studies 230
    sequencing of human genome
        (Human Genome Project) 7,
        112–16, 122–4, 320n, 324n

structure 5–6, 92–3, 181, 185–90, 318
  determination 5–6, 94–7
  double helix 6, 96, 97, 98, 108, 182, 185, 186
  topology (incl. knots) 182, 185–90
  see also mutations
DNA helicase 97
DNA ligase 97
DNA polymerase 97
DNA-space 152–3
DNA topoisomerases 97, 186–7
Dobzhansky, Theodosius 72
dodecahedron 138, 143
dog
  cat seeing a 173–4
  cladogram 128–9
  gait 167
  pedigree breeds, maintaining 228
domains (of life) 22, 33, 124, 125, 137
Douady, Stéphane 50, 51
double helix of DNA 6, 96, 97, 98, 108, 182, 185, 186
dove–hawk game 218, 219
Drosophila melanogaster (fruit fly) 85, 256
dunes, sand 205–8
dynamics
  network 252, 254, 255
  population see populations

E
Earth (the)
  age 56–7, 60, 321n
  aliens visiting 292–4
  Galileo and 14

  intelligent life nowhere other than 292
  life on see life; terrestrial life
  Moon's origin from 333n
  population peaking 264–5
Earth-like life on other planets 291, 312
Earth-like (terrestrial) planets 303
  life on 291, 302, 303
  super-Earths as 313–15
ecology
  food webs 246
  plankton paradox 258, 269–71
ecosystem
  'balance of nature' metaphor 270
  definition 259
  epidemics in special type of 271
edges in networks 253, 254
Edinburgh–Cambridge model of foot-and-mouth disease epidemic 272
egg (ovum)
  development, epigenetic effects 109
  fertilisation 26, 88
Eggenberger Hotz, Peter 211, 212
eighteen-dimensional space 153, 156
elasticity
  energy landscapes for proteins and 193–4
  plant shoot growth and 52–3
electrical activity (nerves) 160, 161–2
  visual cortex 175, 176, 177, 178
electron density, Fourier transform 93
electrophoresis of DNA fragments on gels 113
Elements, Euclid's 138

elephant, African 232
ellipses and viral structure 151
Elmhirst, Toby 136
embryonic development 26–9, 199
endemic disease 259
endosymbiotic theory of organelle
    origin 87
energy landscape, molecular 193–4
environment
    evolution and 3–4, 58, 62, 69
    Lamarckian theory 59
    genetically-modified organisms
        released into 106
enzymes
    DNA replication 97–8
    DNA site-specific cuts (for
        sequencing) 115–16
    in knot theory 189–90
    see also proteins
epidemics of disease 259, 271–3
    foot-and-mouth disease 271–3
    networks in 254–5
epidemiology 259
epigenetics 10, 107–9
equations (and their solutions),
    symmetry 204–5
equilibria see steady states
equines see horse
Ermentrout, Bard 175, 177
Euclidian geometry 138, 139, 140,
    141, 143, 151
eukaryotes 23–6
    cell reproduction see meiosis;
        mitosis; sexual reproduction
    cell structure 23–6
    chromosome 85, 88–90
        recombination 88, 89–90,
            189–90
    domain of 22, 125, 137
    organelle origin 87

Euler, Leonhard
    and Königsberg bridges 252–3
    and polyhedra 144, 146
Euler's formula 144
Europa 308–9
evolution 3–4, 56–76, 124–7,
    213–45, 214, 215, 310, 319,
    321–3n, 328–30n
    alien life vs life on Earth 295–8
    divergence in see species
    evidence for 72–5, 101–2, 221
    game theory 219–20
    genetic code 100–1
    geology and 59–60, 61, 62, 65–7,
        74, 221
    humans see humans
    new species in see species
    origin of life in hydrothermal
        vents 300
    protein folding and 195
    'theory' (meaning of the word) 72
    'Tree of Life' 101, 102, 124–37,
        324n
    universality 310
Evolve 287
excitability, axonal 163–4, 165
exclusion, competitive 258, 269–70
exoplanets 302–3, 308, 313–15
    life on/from see alien life
    plate tectonics 313–15
exponential population growth
    64–5, 267
extelligence 296
extraterrestrial life see alien life
extremophiles 301, 309
    extraterrestrial 309

F
family (taxonomy) 33
family resemblance 224

family trees
humans 126
snapping shrimp 229
feedback loops
between genes and organisms
103, 211–12
habitable planets 308
plant growth 53
feline species, evolutionary
relationships between various
131–2
Fermat spiral 48, 49
fertilisation 26, 88
Fibonacci branch 51
Fibonacci numbers 43–54, 260, 262,
320–1n
figure-of-eight knot 190, 191
finches (*Geospiza*) 239, 244
Darwin's (incl. Gould's
observations) 61, 61–2,
242–4, 244–5, 329n
fish, transition to land animals 75,
295–6
fitness 215
concept of 215, 231
survival of the fittest 69, 70, 214,
215
Fitzhugh–Nagumo equations 163–4,
325n
Fitzroy, Robert 61
five digits (pentadactyly) 295, 296
five-dimensional space 151, 156
flagella 19, 22–3
flames as alien life forms 277–8
flea, Hooke's drawings 20
flour beetle, population dynamics
268–9
flowering plants
classification 37

colour of flower, inheritance
79–84
petal numbers 39–40, 41, 43, 48,
51, 52
flu virus, swine 153
folding of proteins 182, 192–7,
327n
food-crossing airway, evolution
295–6
food web 246
foot-and-mouth disease
epidemic (2001 in UK) 271–3
icosahedral structure of virus 141
forces (growing stem) 49, 52, 53
forest and savannah African
elephants 232
fossils 73–5, 221, 244, 296
Cambrian explosion 134
meteorite 289
four-legged locomotion 165–9
Fourier transform of electron
density 93
fourth (4D) in viral structure
147–53, 156
foxes
on cladogram 129
with rabbits and owls 70–1, 258
fractals 201
fractions
continued 44
Fibonacci 43–4, 47, 49, 50, 51
Franklin, Rosalind 6, 96, 97
free radicals and origin of life 300
frog development 27
'frozen accident theory' 101
fruit fly (*Drosophila melanogaster*) 85,
256
fuchsia petals 39, 51–2
Fuller, Buckminster 139, 145

funnels in energy landscapes for
proteins 195

G

Gaia hypothesis 259
gait 165–9
Galápagos Islands 61, 240–5
Galileo 14, 15, 20, 49, 317
Galton, Francis 78
'Game of Life' (Life) 284–5, 286, 287
game theory 216–20
Gamow, George 98, 99
Ganymede 309
gastrulation 27, 191, 211, 212
Gause, Georgyi 258, 261, 269
'gay' gene 118–19
gel electrophoresis of DNA
fragments 113
gene(s) 4–5, 91, 117–19
on chromosomes 5, 22
on same chromosome close
together 85
disease risk and 118, 122
flow, and speciation 231, 232,
233, 235, 238, 239, 243
jumping 104
in morphogenesis 210, 211–12
passage/transfer between
organisms in evolution 134–6
horizontal gene transfer 135–6,
137
petal numbers and 39
proteins coded by see proteins
regulatory networks 247, 256–7
RNA interference with 108–9
'selfish' 103
simplistic models of mode of
action, cautions with 104–5
'slavish' 103

two distinguishing definitions
117–18
see also alleles; DNA; mutations
gene therapy 122
genera (taxonomy) 2, 33
generative spiral 46, 48–9
genetic code (triplet bases) 98–101
alternatives to standard form
279–80, 299–300
genetic engineering (incl. genetic
modification) 103–7
genetic variation and diversity 89
humans 121
genetics 4–5, 77–84
development and 28–9
see also epigenetics; inheritance
genome 10
as hotbed of dynamic interactions
106–7
human see humans
geodesic dome 145
geology and evolution (historical
perspectives) 59–60, 61, 62,
65–7, 74, 221
geometry 138–57
viral 139–49, 153–7, 325n
gerbil gait 168
germ cells see egg; sperm
Gilbert, Walter 113
GJ 876d (exoplanet) 313
GJ 1214b (exoplanet) 302–3, 313
glasses (spectacles), invention 15–16
Gliese 876 (star) 313
Goldberg polyhedra 147
golden angle 44, 46, 47, 48, 49, 50,
51, 54
golden number 44, 47, 50, 52, 54–5
Goldilocks (habitable) zone 304,
306–9

Golubitsky, Marty 165, 167, 168, 171
Gould, John 61–2, 242–3
Gould, Stephen Jay 134
grains of sand and dune patterns 207, 208
Grant, Peter and Rosemary 244
gravity 311
greenhouse gases and greenhouse warming 307–8
grey aliens 294, 298
Griffith, Frederick 93–4
growth
    plant and shoot tip 44–53
    regulating patterns of 210
guanine (G) 6, 94, 95
    cytosine pairing with 96, 97
    methylcytosine pairing with 108
gull species 222, 223

H
H1N1 swine flu 153
habitable zone, Sun's 304, 306–9
Haeckel, Ernst 126, 127, 128, 136
haemoglobin 192–3, 327n
hair colour, inheritance 88–9
Haldane, JBS 35, 43, 73, 74
hallucinations, visual 171–9
hawk–dove game 218, 219
HD 18973b (exoplanet) 302–3
heartbeat, leech 169–71
heat, planets and the physics of 306–7
helical structure
    double (DNA) 6, 96, 97, 98, 108, 182, 185, 186
    viruses 140, 144
helicase 97
Hennig, Willi 130

heredity *see* gene; genetics; inheritance
Hermite, Charles 139
herring gulls 222
Herschel, Sir John 60
Hershey, Alfred 94
heuristics in cladistics 131, 133
hexagons
    pineapple 41, 53
    viral coat 140, 147, 155
hexamers, viral capsomers 144, 146
Higgs boson 293
Hippocrates 259
Hirudo medicinalis heartbeat 169–71
HK97 phage 157
Hodgkin–Huxley equations 161–2, 326n
Hofmeister, Wilhelm 45–7, 48
hominid evolution 116
Homo sapiens see humans
homosexuality and the 'gay' gene 118–19
honeycomb patterns, perception 172, 175, 177
Hooke, Robert 20–1
Hopf link 190
horizontal gene transfer 135–6, 137
horned beetles, mating strategies 226–7
horse
    evolution 74, 221
    motion/gait 166, 168
Huisman, Jef 270
humans (*Homo sapiens*)
    biped gait 169
    brain 159
    chromosomes 88–9
    evolution 116
        Cambrian explosion and 134
        from non-humans animals 75–6

genome 117, 121–3
  individual variations in
    sequence 121
  sequencing (Human Genome
    Project) 7, 112–16, 122–4,
    320n, 324n
  peak population on Earth 264–5
  taxonomy 35, 36
Humboldt, Alexander von 60
hydrogen bonds between DNA
  bases 96
hydrogen streams on Titan 316–17
hydrogen sulphide and origin of life
  300
hydrothermal vents, life evolving
  around 300

I
ice ages 305
icosahedron 139
  viral 140–7, 154, 156, 157
immunisation (vaccination), foot-
  and-mouth disease 271, 273
Imperial model of foot-and-mouth
  disease epidemic 272–3
independent combination of
  chromosomes 85
infectious disease see disease
influenza virus, swine 153
information theory, digital messages
  in 153
inheritance (heredity) 77–90
  of acquired characteristics 58, 59
  blending theory 78
  see also genetics
instability (loss of stability)
  in speciation, onset of 238, 240
  symmetry-breaking 205–6, 238
intelligent life
  Earth as only planet with 292

evolution 296
extraterrestrial 290
  visiting Earth 292–4
interneurons and leech heartbeat
  170
Interspread model of foot-and-
  mouth disease epidemic 272
invariant, knot 187–8
Io 309
ion leakage, axons 161–2
iron core of planets 314

J
Janssen, Zaccharias (and son Hans)
  16
Jeffreys, Harold 241
Jones polynomial 188
Jorgensen, Richard 108
jump (in four-legged gait) 167, 168
jumping genes 104
'junk' DNA 181
Jupiter
  life on Earth and influence of
    303–4, 305, 310–12
  moons 308–9
Jurassic Park 116

K
Kauffman, Stuart 233, 239, 254,
  278–9
Keef, Tom 156
Kepler, Johannes 16, 66
kinesis 26
kingdoms 2, 33, 124, 125
Kirby, William 60
knot theory 183–91, 327n
Kondo, Shigeru 202
Kondrashov, Alexey and Fyodor 234
Königsberg bridges 252–3

Kunz, Martin 51
Kuramoto, Yoshiki 255
lacewing species 102, 235, 235–6

L

Lack, David 243–4
lakes, Titan's 315
Lamarck and Lamarckism 58–9, 69
land animals, transition from fish to
    75, 295–6
Langton, Chris 286, 287
Large Hadron Collider 293
lateral inhibition (neural
    development) 255–6
lattice patterns and viruses 154–7
leaf arrangement on stem
    (phyllotaxis) 40–1, 43–6, 47, 49,
    50, 54, 55
leech heartbeat 169–71
Leeuwenhoek, Anton van 17–18, 19
legged motion of animals 165–9
lenses (glass), invention 13, 16–17,
    19
    see also microscope; telescope
Leonardo of Pisa (Fibonacci)
    numbers 43–54, 260, 262
leopard, evolutionary relationships
    with other feline species 131–2
Leslie models and matrices 261,
    262–3
Levine, Phoebus 92, 94
Levinthal, Cyrus, and the Levinthal
    paradox 194
Levitov, Leonid 51
Lewis, Julian 255
Lewontin, Richard 9–10
Liddington, Robert 147
Life ('Game of Life') 284–5, 286, 287

life (living things) 275–316, 331–4n
    alien see alien life
    artificial 285, 286–7
    classification see taxonomy and
        classification
    definitions 275–88, 331–2n
    evolution see evolution
    synthetic 288
    'Tree of' 101, 102, 124–37, 324n
    see also species; terrestrial life
ligase, DNA 97
Lightfoot, John 66
limbs, pentadactyl/five digits 295,
    296
Linnaean classification and
    Linnaeus 2–3, 32, 33, 35, 36,
        37, 38, 101, 124, 125, 126, 222
Linnaean Society meeting, Darwin's
    and Wallace's paper work 56,
    58
lizard (North American west coast)
    mating strategies 213–14, 215–16
    taxonomy 224
locomotion, animal 165–9
logarithmic (exponential)
    population growth 64–5, 267
logarithmic spirals 55
logistic (sigmoidal) population
    growth 261–2
    variant model 266–7
loops
    closed see closed loops
    feedback see feedback loops
    in knot theory 184, 185, 186,
        187, 188, 189, 190
Lovelock, James 259
Lucas, Édouard 43
Lucas branch 51
Lucas numbers 51–2, 54

lungs and the evolution of land animals 295–6
Lyell, Charles 60, 62, 68

M

McCarty, Maclyn 94
McClintock, Barbara 104
MacKay, Chris 316
Macleod, Colin 94
mad cow disease (BSE) 193
magnetic fields, planetary 314
Maini, Philip 255
male homosexuality and the 'gay' gene 118–19
Malthus, Thomas 64–5
Mandelbrot, Benoît 201
mantle, Earth's 241, 242, 314
many-celled organisms 21, 22, 26, 28
Mars 276, 304, 305, 307
   life on 289, 306, 312
   life on Earth and influence of 305
mathematical biology see biomathematics
mating see breeding
Mattick, John 108
Maxam, Allan 113
May, Robert 266–8
Maynard Smith, John 219
Mayr, Ernst 222
mechanics
   celestial 311
   gait 166–9
   growing stem forces 49, 52, 53
media images of aliens 297–8
medicine and the Human Genome Project 122
Meinhardt, Hans 201, 209, 210
meiosis 85, 86, 88

men, homosexuality and the 'gay' gene 118–19
Mendel, Gregor 4–5, 77–84, 89, 318, 323n
Mercury 308
Mereschkowski, Konstantin 87
mescaline 171
metallo-peptides and origin of life 300
metaphase 87
meteorite fossils 289
methane
   exoplanetary 302–3
   Titan's lakes of 315
      methane-based life 316
methylated DNA 108
5-methylcytosine 108
Micrographia 20
microscope 1–2, 10, 13, 14–15, 16–20, 21–2, 318
microtubules 24–6, 86–7
Miescher, Friedrich 92
mirror symmetry 204
'missing links' 74–5
mitochondria 23–4, 300
   DNA 87, 100, 130, 323n
   evolution 87
mitosis 86–7
   mitotic spindle 25
Möbius band 183
mockingbirds, Darwin's observations 61–2
mode interaction, neural networks 171
molecular topology 181–97
Monk, Nicholas 255
Moon 305–6
   Galileo and the 14
   life on Earth and the influence of 304

origin 305, 333n
moons of other planets 304
  in Solar System, life on 304,
    308–9, 315, 316
Morgan, Thomas Hunt 85
Morgenstern, Oskar 218
morphology (form), animal
    199–203, 209–12
motion see movement
movement/motion
  animal 165–9
  continental see plate tectonics
  eukaryotic cell 23–5
  prokaryotic cell 22–3
  stripes on animals 202–3
multicellular (many-celled)
    organisms 21, 22, 26, 28
Murray, James 201–2
mutations
  development and 28
  genetic changes arising by
    methods other than 135
  phenotypic change due to 102
  in speciation studies 230, 234,
    235
  viral 153
  see also allele
Mycoplasma laboratorium 288
Mycoplasma mycoides genome, man-
    made (synthetic) 288

N
Nash, John, and Nash equilibrium
    218
natural selection 4, 56, 65, 67–71,
    102, 215, 219, 222, 238, 239,
    243, 255
  energy landscapes for proteins
    and 195

Lamarckian form 59
nautilus shell, spiral pattern 54–5
Nazca plate 241, 242
Neanderthal genome 116
Nectocaris 134
neo-Darwinists 117
nerve cells see neurons
nervous system development 27–8
  networks in 255–7
  see also neuroscience
networks 246–57, 330n
  general theories of structure
    254–5
  neural 165–80, 246, 255–7
neural groove and tube 28
neurons (nerve cells) 159–80
  networks 165–80, 246, 255–7
  single 160–1
  see also axons
neuroscience 158–80, 255–7,
    325–7n
  developmental see nervous system
    development
Newcomb, Simon 148
Newton, Isaac 8, 49, 66
niches and the plankton paradox
    258, 269
Noah's Ark 18–19, 30–2
nodes in networks 253, 254
non-lattice patterns and viruses
    154–7
Notch 256, 257
notochord 33, 134
novelty of maths in biology 8
nucleotides 92, 94
  see also bases
number sequences, plants 38–55

O

oceans on Europa 308–9
O'Connell, Richard J 312
Ohm's law 162
Okazaki fragments 97–8
Onthophagus ventris (horned beetle), mating strategies 226–7
Opabinia 134
order (taxonomy) 2, 33
organelles 23–4
  evolutionary origin 87
  in mitosis 86–7
organic molecules, exoplanetary 302–3
Origin of Species 3, 4, 68–9, 136, 220, 223
oscillations, populations 267, 269, 270
overhand knot 187, 188
ovum *see* egg
owls, competition with foxes 70–1, 258
oxygen
  origin of life and 300
  transport 192–3
    haemoglobin 192

P

Pacific plate 241
Palacios, Antonio 171
Paley, William 60
pangenesis 78–9
parasites, aliens as 297
parochial features of life 295, 296
particle physics, higher dimensions in 152
patenting, biotechnology 107
pattern formation 198–210, 328n
Pauling, Linus 97, 146

payoff (in game theory) 218, 219, 220
  hawk–dove game 218, 219
  scissors–paper–stone 216
peas, Mendel's 4, 77, 79, 80, 82, 85
Penrose tilings 154, 156, 157
pentadactyly (five digits) 295, 296
pentagonal Penrose tiling 154
perception, neural networks 171–8
petal numbers 39–40, 41, 43, 48, 51, 52
peyote 171
phages 5, 94, 140, 157
phenotypes
  mutations causing change in 102
  two new diverging from each other in sympatric speciation 244
phi (ϕ; golden number) 44, 47, 50, 52, 54–5
phyla 2, 33
phyllotaxis (leaf arrangement on stem) 40–1, 43–6, 47, 49, 50, 54, 55, 320–1n
Phylogenetic Systematics 130
Physarum polycephalum see slime mould
physics
  of heat, planets and 306–7
  higher dimensions in 152
pig (swine) flu virus 153
Pigliucci, Massimo 223–4
Pikia 134
pili 22, 23
pineapple, spiral patterns 41, 42, 53
Pinto, Carla 168
Planck's law 306–7
planes and viral structure 150–1
planets, life on other *see* alien life

plankton paradox 258, 269–70, 330–1n

plants
chloroplast evolution 87
flowering *see* flowering plants
Mendel's studies 79–84, 85, 318
number sequences 38–55
taxonomy 37

plasmodia, slime mould 249

plate tectonics (and continental movement) 312–15
Earth 241–2, 312–13
exoplanetary 313–15

Pneumococcus 93–4

Poincaré, Henri 266

'points' in space and viral structure 150–1

pollen spread in genetic modification experiments outdoors 106

polymerase, DNA 97

polynomials, knot theory 188

polyoma virus 154

Pomacanthus imperator (angelfish) stripes 202–3

populations
dynamics (and growth and change over time) 259–70
chaotic 266–71
Malthus on 64–5
peak, humans on Earth 264–5
speciation and 238–9

potassium ion leakage, axons 162

pre-patterns 209

primary-fate cells 256, 257

primates 35
brain 159
see also apes

primordia (shoot tip) 45–53

probability
foot-and-mouth disease epidemic probability model 272
Mendel's studies and 83–4

prokaryotes 22–3
chromosome 85
domain of 22, 125, 137
as origin of eukaryotic organelles 87
reproduction 22, 84, 86
structure 22–3

prophase 87

proteins 181
genes/DNA coding for 5, 98–101
number of proteins per gene 104–5, 122
network of interactions 257
structure 181
folding 182, 192–7, 327n
see also amino acids; enzymes

protists 18

pseudo-icosahedra 139
viruses 147

Punnet square 81, 82

Q

quadruped locomotion 165–9

quasicrystal patterns 154

R

rabbits
competition between, and with foxes 70–1, 258
Fibonacci's 42–3, 260
Leslie models and 262–3

radiation (heat), planets and reflection of 306–7

rail system and slime mould

networks, comparisons 248, 249, 250, 251
random events 8
  in evolution 71–2
  of humanity 134
  in Mendelian genetics 83, 84
random networks 254
Rare Earth (Peter Ward's and David Brownlee's) 290, 293, 303, 304, 309, 312, 314
rat gait 168
recombinase 190
recombination (chromosomal) 88, 89–90, 189–90
  site-specific 189–90
reefs, coral, Darwin on 62, 63
reflection of radiation, planets and 306–7
refractory period (nerve impulse conduction) 163
religious/theological views (historical)
  creation and evolution 60, 62, 63, 64, 66, 67, 75, 76
  solar system 14
replication/self-replication (copying)
  DNA 97–8, 185–6
  life forms 278, 281
  models for 281–6
representation of itself, replicating device containing 281–2
reproduction (replication with errors) 278, 286
  eukaryotic cell *see* meiosis; mitosis; sexual reproduction
  prokaryotes 22, 84, 86
  see also breeding
reproductive isolation 228
resolvase, Tn3 190

resources
  competition for 3, 65, 70, 258, 270
  population growth and, Malthus on 64
ribonucleic acid *see* RNA
ribosomes 23
RNA (ribonucleic acid) 100, 279
  interference (with genes) 108–9
  transfer 279
  viral 140, 279
robots
  replicating (of Von Neumann) 281–2
  Solar System exploration 276
rotational symmetry 204
Rule 110 automaton 285, 287

S
Saccharomyces growth curves 261, 262
saddle points in energy landscapes for proteins 195
safety of genetic modification 105–6
sand dunes 205–8
Sanger DNA sequencing method 113–14
Sasselov, Dimitar 312
satellites, natural *see* Moon; moons
Saturn
  Cassini–Huygens spacecraft mission to 315
  moons 309, 315–16
savannah and forest African elephants 232
Schizoasaccharomyces growth curves 261, 262
scissor–paper-stone game 216, 218
seagull species 222, 223
secondary-fate cells 256, 257

seed patterns 48–9
  sunflower 42, 48–9
selection 70
  natural *see* natural selection
  sexual 214
self-replication *see* replication
'selfish' gene 103
SETI project 276
sex chromosomes 89
sexual orientation, genetic factors
    118–19
sexual reproduction
  bacterial 23
  eukaryotic 26, 84–5, 86
    see also meiosis
sexual selection 214
shape in development 28
shell patterns (form and markings)
    54–5, 201
Shoemaker–Levy 9 (comet) 303, 304
shoot (tip) growth 44–53
shotgun DNA sequencing 113–14,
    116
shrimp, snapping 229–30
side-blotched lizard mating
    strategies 213–14, 215–16
sigmoidal population growth *see*
    logistic population growth
significance of genetic association
    studies 119–20
silicon-based lifeforms 280
single-celled organisms 19, 21, 22
Sinvervo, Barry 213
site-specific recombination 189–90
6^*_2 (knot) 191
six-dimensional spaces 156, 156–7
'slavish' gene 103
slime mould (*Physarum
    polycephalum*) networks 248–50
  mathematical modelling 250–2

Smale, Stephen 266
small world architecture 254
snapping shrimp 229–30
social networking 247–8
sodium ion leakage, axons 162
software, protein folding prediction
    196–7
Solar System (our) 266, 276, 290,
    302, 304, 305, 308, 311–12
  Galileo and the 14
  habitable zone 304, 306–9
  Jupiter's absence from, effect
    311–12
  planets outside of *see* exoplanets
  robotic exploration 276
somite patterning 209, 210
space, viral geometry and
    dimensions in 147–57
species
  brain size comparisons between
    158–9
  competition within and between
    see competition
  definitions 222–4
  estimates of total number 35
  genetic change distinguishing two
    species 101–2
  new, formation (speciation -
    divergence or branching from
    one into two or more) 124,
    126–8, 131–2, 136, 223,
    228–40, 244–5
    humans from chimpanzees 293
  origin *see* evolution
  in taxonomy 2–3, 33
spectacles, invention 15–16
Spencer, Herbert 69, 70
sperm 26, 88
spikes, voltage 162, 163, 164
Spina, Aleesandro 16

spinal cord development 28
spiral patterns
  nautilus shell 54–5
  plants 41–2, 46, 48–9, 50, 51, 52,
    53
  visual perception 172, 176, 177
  see also double helix
spontaneous generation 20
spots (patterns) 199, 201, 202, 328n
stability
  loss see instability
  population 270
star(s), habitable zones 304, 306–9
  see also Sun
starfish symmetry 204
START sequence (triplet base) 99,
  100
statistics, correlation studies in
  genetics 119–20
steady states (equilibria) 265
  populations 267, 270
Steiner spanning tree 251, 252
stem
  growth 44–53
  leaf arrangement on (phyllotaxis)
    40–1, 43–6, 47, 49, 50, 54, 55
STOP sequences (triplet base) 99,
  100
Streptococcus pneumoniae
  (Pneumococcus) 93–4
stripes (patterns)
  on animals 36, 37, 199, 201, 202,
    328n
    moving 202–3
  of neural activity in cortex 176,
    177
  on petals 108
  sand dunes 205, 206, 207
Strobel, Darell 316
strong alife 287

Sun
  Galileo and the 14
  habitable zone 304, 306–9
  see also stars
sunflower seed patterns 42, 48–9
super-Earths 313–15
survival
  of the fittest 69, 70, 214, 215
  game theory and strategies for
    219
Sutton, Walter 85
Swain, Mark 302
swine flu virus 153
symbiotic theory of organelle origin
  87
symmetry 203–9
  breaking 203–9
    in neural network development
      257
    in speciation 236–8, 240, 244,
      245
    in scissor–paper-stone game 216,
      218
  transformations see
    transformations
  in viral geometry 142–3, 147
sympatric speciation 228, 230, 231,
  233, 234, 235, 236, 237, 238,
  240, 244, 245
  onset of instability 238
synthetic life 288
Szybalski's rule 95–6

T
T2 phage 94
T4 phage 140, 141
TAA (STOP) sequence 99, 100
TAG (STOP) sequence 99, 100
tail markings 202
Tait, Peter Guthrie 47

tandem repeats, variable 121
tangles 189–91
    DNA 182, 189–90
    proteins 182
tautology, evolutionary theory as
        69–70
taxonomy and classification 2–3,
        32–7, 320n
    horse evolution 74
    lizards (North American west
        coast) 224
    'Tree of Life' 101, 102, 126, 127,
        128, 134, 135–7, 324n
tectonic plates *see* plate tectonics
telescope 13, 14, 16–17
temperature, planetary, life and
        306, 307, 308
Tero, Atsushi 248
terrestrial life
    alien life resembling 290, 291
    alien life totally unlike 291
terrestrial planets *see* Earth-like
        (terrestrial) planets
TGA (START) sequence 99, 100
Theia 333n
Theobold, Douglas 137
theological views *see* religious/
        theological views
Thompson, D'Arcy Wentworth 9,
        47, 50, 54
three-dimensional lattice patterns
        155
thymine (T) 6, 94, 95
    adenine pairing with 96, 97
Tierra 287
tigers 36, 37, 109, 198, 199, 237
    evolutionary relationships with
        other feline species 131–2
    stripes 199, 202
tiling theory, viral 154–7

Time Machine (HG Wells) 147–8
tit classification 33–4
Titan 309, 315, 316
Tn3 resolvase 190
tobacco mosaic virus 140
Tokyo rail system and slime mould
        networks, comparisons 248,
        249, 250, 251, 252
topoisomerases 97, 186–7
topology, molecular 181–97
train system and slime mould
        networks, comparisons 248,
        249, 250, 251
transfer RNA 279
transformations
    knot theory 187
    symmetry 205, 207, 237
    virus 142–3
transitional forms in evolution
        ('missing links') 74–5
transmutation of species, Darwin's
        63, 64
transposon 104
transposon Tn3 resolvase 190
transverse sand dunes 206–8
'Tree of Life' 101, 102, 124–37,
        324n
Tribolium castaneum, population
        dynamics 268–9
triplet code *see* genetic code
trivial tangle 189, 190
truncated icosahedron 139, 142
tubes in networks
    gravitational 311
    slime mould 248, 249, 250
    mathematical models 251
tubulin 24–6
Turing, Alan 199
    pattern equations 199–200, 201,
        202, 203, 209, 210

universal Turing machine 285
Twarock, Reidun 147, 154, 156
twin studies 118
two dimensions, lattice patterns in
155

U

Ulam, Stanislow 282
uncertainty 8
unicellular (single-celled) organisms
19, 21, 22
Unitarianism, Darwin family's 63,
67
universal common ancestry 58, 137
universal features of life 295, 296,
297
universe, symmetry 204–5
unknot 188, 190, 191
Ussher, James 66
Uta stansburiana (side-blotched
lizard) mating strategies
213–14, 215–16

V

vaccination, foot-and-mouth disease
271, 273
Valencia, Diana 312
van Beneden, Edouard 25
van Leeuwenhoek, Anton 17–18, 19
variable tandem repeats 121
variety of maths in biology 8
Venter, Craig (and the Craig Venter
Institute) 112, 121, 288
see also Celera genomics
Venus 306
in habitable zone 307
viruses
bacterial (=phage) 5, 94, 140, 157
geometry 139–49, 153–7, 325n

RNA 140, 279
visual system 171–80
Vogel, Helmut 48–9
voltages (axons)
leakage 162
spikes 162, 163, 164
Von Neumann, John 287
game theory 217–18
replicating devices 281–4

W

Wächtershäuser, Günter 300
Wallace, Alfred Russel 56, 67–8, 68
Wallin, Ivor 87
Ward and Brownlee's *Rare Earth*
290, 293, 303, 304, 309, 312,
314
water (planetary)
exoplanetary 303, 313–14
liquid 304, 308, 312
vapour
Earth 307, 308
exoplanetary 303
Watson and Crick 5–6, 7, 96, 97,
98, 101
waves
animal markings or patterns and
202, 209–10
shoot growth and numbers and
patterns of 53
weak alife 287
Wegener, Alfred 241
weight, body, and brain weight
158–9
Weissing, Franz 270
Weizenbaum, Joseph 317
Wells, HG, *Time Machine* 147–8
Whitehead link 187, 190, 191
Wilkins, Maurice 6, 96, 97
Williams, Lance 179

Wilson, Hugh, and Wilson–Cowan
    equation 175
Wittgenstein, Ludwig 224
Wolfram, Stephen 285
Wolynes, Peter 193
Wonderful Life 134

X
X chromosome 89
  'gay' gene 118–19
X-ray diffraction studies 92–3, 319
  DNA 93, 96, 186
xenobiology and xenoscience 291

Y
Y chromosome 89
yeast
  growth curves 261, 262
    network of protein interactions
      257

Z
zebra stripes 36, 37, 198, 199, 202
Zoonomia 57, 58
Zweck, John 179